Ageless Quest

One Scientist's Search
for Genes That Prolong Youth

Ageless Quest

One Scientist's Search
for Genes That Prolong Youth

Lenny Guarente
Massachusetts Institute of Technology

COLD SPRING HARBOR LABORATORY PRESS
Cold Spring Harbor, New York

AGELESS QUEST
One Scientist's Search for Genes That Prolong Youth

Publisher	John Inglis
Acquisition Editor	Alexander Gann
Developmental Editor	Jody Tresidder
Project Coordinator	Joan Ebert
Production Manager	Denise Weiss
Production Editor	Patricia Barker
Cover Designer	Ed Atkeson
Desktop Editor	Danny de Bruin

Front Cover
Actress Carole Lombard, ca. 1937 (photographer Eugene Robert Richee)
© Hulton-Deutsch Collection/CORBIS

Diagram Credits
The following art has been redrawn with permission: p. 93 from Frye 2000 (*Biochem. Biophys. Res. Commun.* **273**: 796 [© Academic Press]); p. 111 from Dutnall and Pillus 2001 (*Cell* **105**: 163 [© Cell Press]).

Library of Congress Cataloging-in-Publication Data
Guarente, Leonard.
    Ageless quest : one scientist's search for genes that prolong youth / Lenny Guarente
         p. cm.
Includes bibliographical references and index.
    ISBN 0-87969-652-4 (cloth : alk. paper)
    1. Aging--Genetic asects. 2. Aging--Molecular aspects. I. Title.
QH608 .G83 2002
571.8'78--dc21
                                                                                        2002031513

10 9 8 7 6 5 4 3 2 1

All Cold Spring Harbor Laboratory Press publications may be ordered directly from Cold Spring Harbor Laboratory Press, 500 Sunnyside Boulevard, Woodbury, New York 11797-2924. Phone: 1-800-843-4388 in Continental U.S. and Canada. All other locations: (516) 422-4100. FAX: (516) 422-4097. E-mail: cshpress@cshl.edu. For a complete catalog of all Cold Spring Harbor Laboratory Press publications, visit our World Wide Web Site http://www.cshlpress.com

For Jef

Contents

Preface

WHY WRITE A BOOK ON AGING and make it a personal account at that? The simple answer is that there is a story to tell. It is not a story that summarizes everything we are beginning to understand about aging, or even one that covers a wide swath of current research on the topic. For these omissions, I apologize to those scientists whose work is not given its well-deserved due here. Instead, this book is a personal journey that began some fifteen years ago.

The questions of how and why we age have fascinated people for millennia. The aging process is extremely complex with many causes. It appears to proceed inexorably, like the waning of daylight on a summer evening. But things are not always as they seem. Nature has very cleverly evolved a way to put the brakes on aging in times of scarcity. Why would such a survival mechanism be advantageous? What would be the point in evolutionary terms? Imagine, as an example, I had a gene to slow down the aging process during a famine and you didn't. When the famine is over, and assuming I survived, I would likely remain young enough to reproduce and you would not. In that way, the gene which slowed down aging will predominate in subsequent generations.

Interestingly, it has been shown that aging can be slowed down in rodents subjected to an artificial famine-like situation—a low-calorie diet. Perhaps this is a demonstration of such a survival mechanism. It is the issue of how survival mechanisms might work that has spurred my research and, not surprisingly, which drives this narrative.

Research usually follows a twisted, turning path, inevitably involving stumbles and blind alleys. Discovery comes only gradually. In this particular journey we are probably not yet out of the labyrinth. But a picture has already come into focus that, amazingly, originates in a study of simple yeast cells and may have implications for humans.

Even based on current knowledge, the new findings suggest the existence of a universal survival mechanism that is relatively simple. It may be possible to find drugs to extend the human life span. I discuss the tale of how a company was started on the basis of this premise. The chronicle of our research is by no means the final word on aging. It is intended as an adventure story that moves through the biological world to arrive at interesting, general conclusions. The narrative also focuses on the process of doing scientific research, and, most importantly, on the young scientists who have moved the frontier forward. I try to convey how new findings are made, while acknowledging the sometimes chance nature of scientific discovery.

There is a glossary of useful scientific terms at the end of the book. The definitions should help the nonspecialist reader.

I thank many who have been important influences leading to this book. My mother realized the importance of education long before I did. My many colleagues in the research community at MIT and around the world have inspired and taught me. Jana Koubova helped me develop the flow of the book and read every word with care. Jean Monroe provided important editorial guidance in the book's early stages. Alex Gann encouraged me when I was unsure of what I had done. Mark Ptashne provided many useful suggestions on the organization of the book. Jody Tresidder was a wonderful editor and supporter. Finally and most importantly, I thank the young scientists who have blessed my lab over the years. These fascinating individuals not only generated the discoveries, but also made life always interesting, exciting, and fun.

Lenny Guarente
August, 2002

CHAPTER 1

··

Mid-life Crisis

I WAS EXPERIENCING A PREMATURE MID-LIFE CRISIS—recently divorced from my wife and even more recently freed from a tempestuous relationship with a colleague. On the bright side, I had just been awarded joint custody of my 3-year-old son and tenure in the prestigious Biology Department at MIT. The drive for tenure was not an innocent bystander in the unraveling of my personal life. Its aftermath left me pondering what should come next for a 34-year-old molecular biologist whose life to date had been spent reaching for the next rung of the academic ladder.

Striving for tenure at a place like MIT is a special kind of torture. When tenure is granted you don't feel so much that you have accomplished something as that you have dodged a precisely guided bullet. The top universities like to publicize the preeminence of the teaching mission of their faculty. And why not, since every undergraduate pays in excess of $30,000 per year for the privilege of coming and being inspired. But when you as an untenured professor are facing the crucial decision about your future, you know that you will be evaluated mostly on the basis of your research triumphs during your five-year sojourn as an assistant professor.

The climax occurs at a meeting of your senior colleagues. The most important criteria under scrutiny are letters from 10 to 20 of the leading experts in your field, solicited by your department chairman from around the world. This last measure weighs so heavily on your mind that you search through the mental files of all the people you have met at scientific conferences over the years. Did you impress your senior colleagues? Did you make any enemies? The kiss of death is when they say they have never heard of you. The benediction is when they describe you as a leader in your age group. Therefore, most young assistant professors choose to work in an area that is well established and guarantees that a hard day's work will yield a commensurate level of accomplishment. You trust five years of this will yield tenure. By this logic, risk-taking is to be avoided, or at least postponed, until a permanent position has been secured.

With this in mind, I had launched my career as an assistant professor in 1981 by studying a simple, basic problem in biology: how genes are regulated; i.e., how they get turned on and off in cells. My branch of biology, molecular biology, is an experimental science driven by the scientific method. We form hypotheses about how biological processes work and then devise experiments to test our theories. We use the most sophisticated experimental tools available, including, of course, gene cloning and DNA sequencing.

The field of gene regulation was a crowded one, with scientists falling over each other to carry out quite similar experiments to establish rather closely related principles. The good news was that meant one could do experiments quickly, get the results published, become known to the worldwide experts, attract grants to support more experiments, and, in short, follow a safe pathway toward tenure. The bad news was that it was difficult to make a fundamental contribution to science; that is, to make a discovery that wouldn't have been made without you. Clever new discoveries were at best months ahead of similar findings from other labs, or at worst banished to the dustbin because a competitor scooped you by publishing first.

So, after five years of studying genes and wearing the new mantle of tenure, I began to imagine what kind of research would really make a difference. Any risk seemed manageable, since my MIT masters could no longer kick me out the door. For about a year, I thought about infectious disease. The AIDS catastrophe was becoming evident and the causative agent had just been identified, even though a cure did not look imminent. However, an eminent virologist I consulted cautioned me that there was already a huge effort being mounted by the government and assured me that the biotech industry would develop treatments; thus, my efforts would be a drop in the ocean. This sounded like the gene regulation field all over again.

Another area that had always fascinated me was the human brain. How do we learn? What is the biological basis of memory or even consciousness? Surely, this was a problem with a high potential for reward, albeit one with an equal level of risk. The difficulty here was figuring a way to begin to tackle such daunting questions. How one trillion brain cells organize themselves for these mental tasks seemed much too complicated for an experimental approach that would yield a molecular understanding. I had no clue as to how to get started.

Several years ticked by as these deliberations continued. Our research on gene regulation during this time, if anything, became more exciting, but always within the overcrowded arena of that field of biology. Then, I was aided by the fateful appearance of two graduate students who entered the MIT biology department in 1990. Graduate students in biology earn their Ph.D. degrees by carrying out research in the laboratories of biology professors. These students really constitute the engine that moves biomedical research forward. Entering students at MIT spend most of their first year taking classes and choosing the lab in which to conduct their thesis work. The students therefore spend a considerable amount of time talking with professors as a part of this decision process. The two talented, eager students brought to my doorstep by a happy coincidence in that particular year were Brian Kennedy and Nick Austriaco. Brian made a strong initial impression by his physical presence. He was a hulking figure marked by a vivid scar running along one side of his face. Upon first seeing him, I vaguely recalled hearing that there were two students in the entering class, a married couple, who had deferred admission for a year because of a serious car accident. Brian had almost lost his life and spent a year in therapy to regain full use of his body and mind, while his wife was less badly injured. Indeed, it was Brian's extreme intolerance for any more medical intervention that led him to accept his physical scars as nothing more than a minor inconvenience. He had graduated from a selective program integrating math and science at Northwestern University and possessed a keen intellect and logical bent that would allow him to connect the dots in the biological puzzles he would face in the lab. He was calm and self-effacing; a demeanor that concealed a self-confidence and intensity that would become apparent during his research tenure.

Nick Austriaco was a foreign student who was born in the Philippines and grew up in Thailand. His approach to solving problems was different from Brian's and, indeed, from the norm. This first became evident through his unusual but effective solutions to exam questions in genetics class. In subsequent years, the teaching staff would ask Nick to solve new exam questions to prepare them for unusual but correct answers from students. Nick's originality was an essential ingredient to the success that he and Brian had as a team in the initial stages of their research. He also possessed a spiritual side that led him to ponder deeper, less answerable questions than those he would face at the laboratory bench. In this respect he stood quite apart from the many scientists who like to believe that they view life through a purely logical lens. In fact, several years later, after obtaining his Ph.D.,

Nick left a prominent position as an independent scientist at the Ludwig Institute in London to join the priesthood as a Franciscan. Nick and Brian scheduled a meeting with me in January 1991. This was somewhat unusual in that it was not the typical one-on-one meeting between the professor and a student searching for a lab. It was also unusual in another respect. Discussions with students usually detail specific projects that are ongoing or under development in the lab. In this case, our discussion ranged far wider than normal and was influenced by my deliberations about new directions for the lab. I tried to provoke them intellectually by asking whether they could identify any problems in biology that were profoundly important but essentially unsolved. Nick then mentioned a paper he had written in college, which described published research on human cells taken by biopsy and grown in culture. Leonard Hayflick, now at the University of California, had shown in the 1960s that such cells divided only a fixed number of times in a petri dish bathed by growth medium and then entered a dormant state termed cellular senescence. This property of cells has been termed the "Hayflick limit." Many had wondered whether this process provided an important clue about what caused the aging of an intact organism.

As we pondered this question, I realized that I had never really viewed aging as a problem that lent itself to analysis at all. Rather, it seemed like something that just happened, like the erosion of a beachfront or the expansion of the universe. You grow old and, therefore, inevitably you die. The evolutionary biologists had been thinking about aging for decades, but although the subject had certainly reached the domain of the experimental scientist, it was not studied intensely by molecular biologists. In the view of population biologists, the aging of the individuals of a species was best defined by the ever-increasing probability of mortality with age. That is, as we get older our mortality rates shoot up, and the number of years it takes for these rates to double is ever shrinking. More simplistically, I viewed aging as a process that was obvious by outward appearance; one could scan a roomful of people and assess the age of each individual at a glance, because the changes that occur are so stereotypical. Holding to the usual standard of the lab, I raised the objection that we could only study this problem if we could understand it at the molecular level.

I knew that the most incisive way problems of this sort could be approached was by the identification of specific genes that are of key importance in the process. This molecular-genetic approach is well known, for example, in having identified single genes that predispose humans to particular cancers or inherited diseases. Once specific genes have been identified, an understanding of the underlying processes that they affect is usually not far behind. The genes lead to the analysis of their products, which are specific proteins or RNA molecules.

Aging, however, seemed different. Some biologists insisted that aging was more like the simultaneous occurrence of many, many late-onset diseases. This would mean that the culprit genes could number in the thousands, condemning to likely failure any search for one or two genes that were of primary importance. This crucial consideration underscored the risk of such a venture—it could be impossible to complete during my lifetime. On the other hand, the expert biologists could be wrong, in which case the armaments of the molecular geneticist could have the power to pinpoint the key gene that would be the Holy Grail. Intoxicated by this thought, we agreed to read up on the subject and meet again.

At the next meeting, things began to get really interesting. We discovered in the interim that there existed a literature on aging in yeast. Yes, I speak of the yeast used to make bread, beer, wine, and various spirits. Yeast cultures actually consist of live, dividing cells, and these cells produce the carbon dioxide gas that makes bread rise and the alcohol that enhances beverages which have been a staple of every known civilization. Geneticists and molecular biologists had discovered years before that yeast was an excellent organism to study in the laboratory. The cells are easy to grow and, indeed, grow very rapidly. Most importantly, it turns out that many biological processes that occur in yeast are faithfully conserved in animal cells. Over and over, discoveries made in yeast cells have presaged how processes would be found to occur in animals. A striking example of this is a process termed the cell division cycle, which goes awry in many human cancers. Pioneering studies in this area earned two yeast biologists, Lee Hartwell and Paul Nurse, the Nobel Prize in Medicine in 2001. More mundanely, my lab—like many others—had been using yeast to study gene regulation for years. This meant that if there were a large psychological barrier in changing fields from gene expression to aging, at least there would be no impediment posed by a new and foreign experimental system. These new developments were more than a little agreeable to me.

But how do yeast cells age? It was well known that the kind of yeast used for baking and brewing produces more cells by a process called budding, in which the progeny or daughter cell protrudes from the mother cell. The daughter can be distinguished from the mother at birth because it is smaller. It then commences a robust period of growth in preparation to bud off its own daughter at its next division. Meanwhile, the original mother continues to bud off more daughter cells. This process can be followed in the light microscope, and the cells so viewed can be moved around with a needle on a solid support of agar, an extract from seaweed that has the consistency of Jell-O, and growth medium. The fortified agar fills a petri plate, the shallow-lidded plastic saucer that has been a companion of biologists for many years.

In the 1950s, Bob Mortimer and colleagues made a seminal observation. They asked whether the same mother cell could bud off only a specific number of new daughter cells. To measure this number, they took a row of mother cells on a petri plate and moved away their daughters after each cell division. They found that after budding off 15–20 daughters, mother cells apparently slowed their cell division, then ceased dividing entirely, and finally died. When they plotted how many mother cells were still alive (the Y axis of the figure below) versus how many times they had already budded (X axis), they made a startling finding: There was a precipitous decline in the capacity to produce another daughter after around 20 divisions. The average and maximum life spans of any organism are clearly evident from such a survival curve. In this case, we see that yeast mother cells exhibit a feature that is a hallmark of human aging; their mortality rate is ever increasing with age. The only difference, of course, is that in yeast the metric of life span is the number of divisions mothers have undergone and not chronological age.

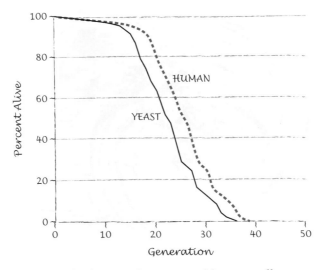

Survival curves in yeast and human cells.

This was very exciting, and we thought we would be able to reproduce the phenomenon in our lab. But if aging were really as complicated as it appeared in higher organisms, would many, many genes be involved in the determination of life span? We had to hope for the existence of a simple mechanism, or our research task would be impossibly complex. We were encouraged by the demonstration that a single gene mutation could extend the life span in the roundworm. This mutation, termed age-1, was isolated in the 1980s by Michael Klass at the University of Houston and studied intensively by Tom Johnson at the Universities of California and Colorado. It thus seemed possible that nature evolved genes to set the pace of aging.

Perhaps genes that could slow aging in times of food scarcity would provide a selective advantage. As mentioned in the preface, imagine if I could survive a famine for much longer than you. Only I would be around to reproduce if bountiful conditions returned. Therefore, those with my survival genes would have a selective advantage during the boom-and-bust cycle that likely occurred during evolution. A survival mechanism of this kind could be relatively simple, and the number of genes that determine it rather few.

Any concerns about numbers of genes, however, soon melted under the zeal that Brian and Nick showed for the prospect of studying yeast aging in my lab. For my part, this also seemed like the ideal recruiting ploy to guarantee that they would select my lab over others around the biology department. Surely, it had to be a good thing to attract two bright, eager, young students to the lab! We thus struck a deal. They could come to my lab to study aging on one condition. They had one year to show some progress or they would agree to switch to one of the more standard projects in the lab on gene expression. I thought that one year would be enough to determine whether the reports in the literature about the way yeast age were correct. If so, it seemed possible that we would somehow find a way to identify the key genes—if the process were that simple.

....

CHAPTER 2

Getting Started in Aging

ROUND THIS TIME, I ran into my department chairman in the hallway and told him that I was planning to start a project to study aging. His immediate response was "You're gonna what?!" I took this to mean he was distinctly underwhelmed by the idea. Several years later we had a similar exchange in the hallway. This time I told him that I was so excited about the aging research that I planned to switch over my entire lab to study aging. His response—"You're gonna what?!" Consistency in mentoring is supposed to make the person who is advised feel safe and secure. In this instance, the effect was quite the opposite.

But my courage was restored by Brian and Nick, who arrived in the lab in June, 1991, liberated from the classroom work of their first year at MIT and loaded for bear. I pulled out a few yeast strains for them to play with in order to acquaint them with the microscopes and the dexterity required for moving cells around with a needle. We were immediately confronted with a practical problem, the solution to which would shape the spirit of cooperation for years to come. Yeast mother cells produce a new daughter cell roughly every 1–2 hours. When they become older, this process slows down and can take up to 3–4 hours. Thus, to follow a life span of 20 mother cell divisions would take about 40–60 hours. The only way to carry out this analysis was for Nick and Brian to work 12-hour shifts and for each to pick up the experiment where the other one left off. The 12-hour shifts eventually began to grind the boys down, however, motivating us to test whether cells could be refrigerated overnight to suspend cell division, and then incubated at warm temperatures again in the morning to resume growth. Mercifully, this protocol gave results indistinguishable from the continuous experiment and was adopted with gusto.

We soon began to observe mortality curves that matched those described in the literature. We also noticed that the mother cells divided more slowly as they aged and gradually assumed a bloated and wrinkled appearance as they approached senescence. Yeast aging was surprisingly similar to what happens closer to home!

Our studies provided an additional bonus. My lab typically worked with several different strains of yeast, which had been propagated separately in other labs for several decades and therefore had evolved genetic differences from one another. These genetic differences endowed the strains with many different properties both in their physical appearances and in their ability to grow under various conditions, thus making one or another most suitable for any particular study. Brian and Nick, in their thoroughness, decided to test all the strains and found that they gave mortality curves of similar shapes, but with vastly different life spans. This immediately suggested that life span in yeast may be genetically determined, but it did not tell us how many genes might be involved in the determination. We continued to hope the number was low.

If one or a few genes determined yeast life span, how would one go about finding them? The currency of molecular geneticists is the creation of genetic variants, or mutants, which alter only one gene and produce an observable change in the organism. In this case, we hoped to generate a mutation in just one of the roughly 6000 genes in the yeast genome that would alter the life span compared to the

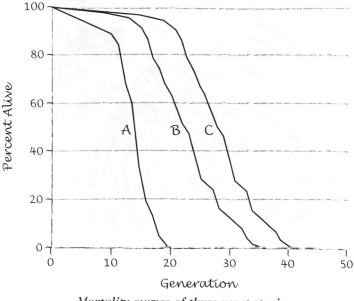

Mortality curves of three yeast strains.

parental strain. But we recognized that mutations that caused a *longer* life span would be more informative than mutations that shortened it. Think of the many genetically determined diseases in people that cause early death, like cystic fibrosis, but which, of course, shed no light whatsoever on the normal aging process. However, if we could identify a genetically isolated group of people who lived extra long, we would expect to glean secrets relevant to normal aging by looking at their genes. Therefore, conducting a genetic screen for mutations in single genes that extended the life span became our obsession.

Only one glitch remained. How could we actually identify this mutant? Imagine that we expose yeast cells to a chemical to generate mutations randomly across the genome and then plate them so that each will grow into a colony. We could ensure the chemical treatment was such that each colony would carry on average one mutation. But to identify a single key gene that determines the life span, we would have to carry out life span assays on cells from more than 6,000 colonies to find the single relevant mutant. Each assay involves the microscopic maneuvering of 20–30 daughter cells away from each mother over a two-week period. Finding the needle in the haystack was obviously too difficult, even for the dynamic duo of Brian and Nick, since a typical experiment could assay several strains, at most. Even if we could stretch to screen 6 different potential mutant strains in each experiment, it would take 40 years to screen through the 6,000 colonies. This would only be feasible if we were already in possession of the fountain of youth!

We therefore put this approach on hold for further consideration. Instead, we contented ourselves with conducting genetic crosses between long-lived and short-lived strains to see whether that would shed light on the genetic basis of their different life spans. Not surprisingly, the progeny had life spans all over the map, ranging from very short to long. This complex result confirmed that the separately maintained laboratory strains had evolved differences in many genes that somehow contributed to their different life spans. Some or all of these genes, like the cystic fibrosis gene in humans, may simply make certain strains sick, and these genes would not provide any special insight into normal aging. In the face of this severe limitation, we shelved our initial plan to sort out all of the genetic differences between these strains and left the years of nature's intricate workings in their blissfully scrambled state.

Instead, serendipity made its first appearance and led us to believe that a direct screen for mutants now seemed practical. Nick attempted to grow cultures of these various progeny strains from samples that had been stored on petri plates in his fridge for a few months. By this time, the agar on the plates had dried out to become wafer-thin, creating a stressful environment for the colonies during the period of storage. Only some of the strains grew when added to fresh media; the others were already dead. Unexpectedly, there was a pattern to which strains grew and which did not. The strains which grew were those that we already knew had the longest life spans!

This showed that the ability of these strains to survive storage, remarkably, correlated with their long life spans, and suggested how we could look for mutants. We would start with a short-lived strain and look for single-gene mutants that did not die when subjected to stress during long-term storage. It would be easy to screen thousands and thousands of colonies in this way, ensuring that mutations across most of the genome would be sampled. Only mutants that passed this first test would be subjected to the more laborious life span assay, greatly simplifying the screening process for mutations that confer longevity.

We thus treated cells with the chemical ethyl methane sulfonate (EMS) which modifies certain of the nucleotide building blocks, A (adenine), G (guanine), C (cytosine), and T (thymine), that make up DNA (deoxyribonucleic acid), and thereby generates mutations. This chemical must be handled with care; experimenters don gloves inside a fume hood that sucks up the exhaust from the lab and releases it outside. The hood provides an additional benefit, since the procedure generates a pungent odor, like an old egg salad sandwich, that is best left unsmelled. The treated cells were spread on petri plates to grow into thousands of colonies, each of which had, on average, one mutation. These colonies were transferred to another plate lacking nutrients to create a stressful environment, and then put in storage for a while. Most of the transferred colonies died during storage (just as the parent strain did). The rare survivors, which appeared at a frequency of about one in a thousand colonies, could be brought back to life by transferring them to a plate with plenty of food. This scheme worked quite well, and for the duration of 1992 and 1993, Nick generated a large collection of yeast mutants that survived the period of stress better than the parental strain.

These stress-resistant mutants were tested one by one in the life span assay. Importantly, when examined under the microscope, about 5–10% of the mutants turned out to have a longer life span. This was the subset of mutants we were seeking, which was now whittled down to only four different mutant strains. One of the mutants proved difficult to study and was discarded, and two of the others were analyzed for the duration of Nick's thesis work. We still do not fully understand why these two mutants have a long life span, a sobering reminder of the snail's pace at which research sometimes progresses. But the fourth and most interesting mutant had the most dramatic effect on life span, a 50% extension. This would correspond to a population of people who routinely lived past 100 years. This mutant had an additional unusual property. It had lost the ability to mate. This unusual coincidence of longevity and sterility was enough to rouse the curiosity of Brian, who assumed the analysis of this interesting strain. Little did we know that the decision to study this mutant would be critical.

CHAPTER 3

First Genetic Insight

RIAN AND I BECAME CLOSE COLLABORATORS during this time, and not just in the lab. Brian loved the game of golf and was a very good player indeed. I had dabbled in the sport in high school, but never became competent. Where I came from, you were much more likely to be proficient in stickball or street-fighting than golf, tennis, or fly-fishing. Golfing with Brian reacquainted me with the fascinating complexity of the game. I am sometimes asked how long I would like to live if aging were to be defeated. I answer 500,000 years—the amount of time it would take to perfect my golf swing.

I grew up in Revere, Massachusetts, a gritty, working-class town with a nature-blessed beachfront accompanied by decaying amusement rides. I was a precocious child by local standards—I quit smoking in the third grade. My father, Leonard Sr., held a lifelong clerical job at the General Electric plant in Lynn, Massachusetts, and my mother, Norma, was of the stay-at-home variety, at least for a while. My younger brother, Dennis, and I were caught in an odd kind of family dynamics. Mom constantly complained that my dad did not work hard enough or earn enough money. She was occasionally forced to borrow money from her parents to make ends meet. Since my grandfather, who emigrated from Italy at the turn of the century, was an industrious restaurateur, he could well oblige these requests. Dad periodically attempted to placate Mom by working a second job, for example, as a clerk in a liquor store, but his efforts were invariably short-lived. It was not surprising that I bought into the litany of Mom's grievances, as she was such an overpowering personality in the family.

At an early age I became intrigued with collecting money, usually in the form of pennies or nickels given to me as presents. When I was four, I got a red plastic piggy bank for Christmas that became my favorite plaything. "Piggy" was soon endowed with a personality of his own; I provided his "voice" and entertained my brother and myself every night at bedtime. A high point was periodically pouring the coins out of Piggy to tally the latest count.

My extended family on Mom's side was very large, and family gatherings were the signature social event of growing up. Unfortunately, the insular nature of this closely knit Italian family brought suspicion on anyone outside. By this family code, friends were given lower status than blood relatives. Furthermore, the rather central role of food and exact meal times heightened the isolation from the outside world. I remember one summer my baseball practice conflicted with dinnertime, creating a clash of wills between my coach and my Mom. I did not complete the season with my team.

However, I was very good at finding solo activities that brought me immense pleasure. I particularly remember discovering pop music when I was eight years old. I would place a pillow under the speaker of the radio—a part of a 1950s style TV/radio/record player console that stood on legs in the living room—and listen for hours. The fact that the TV and record player parts of the console did not function at all was quite irrelevant. My favorite time was when the new top-20 songs of the week were played from number 20 to number 1. Summing up over all the years I have been listening to popular music, I would estimate that I am familiar with about 10,000 songs. This could well be more than the number of words that I know!

My elementary school was down the street, and I got a perhaps surprisingly good education in the first six grades. My teachers were invariably women who practiced their profession with dedication and skill. An underappreciated benefit of the glass ceiling women then faced in many professions was that grade schools, even in working-class communities, were well stocked with talented teachers. In the eighth grade I scored especially well on standardized tests and my math teacher pulled Mom aside and suggested a private high school. Mom quickly seized this opportunity, and when I passed the entrance exams, she decided to go to work as a waitress to pay the tuition. I began contributing to the costs upon turning 16, when I wangled a job at the local grocery store.

Boston College High School was a Jesuit-run institution for smart, but not necessarily wealthy, mostly Irish Catholic kids. It made new demands on me that would bring greater discipline than I had previously known; a daily commute of three hours and a rigorous load of classes. I discovered that hard study was faithfully rewarded with good grades, and by the end of freshman year, I was in the top ten percent of my class. By the end of sophomore year I was in the top three percent, and by senior year I ranked number one in the class. I enjoyed classes that ranged from Latin to physics, and found myself gravitating toward the sciences, but also with a strong interest in foreign languages. Many of the smart kids at school studied classics and not science, and a bunch of them ended up going to Harvard. Perhaps because of this peer pressure, I was more than a little disappointed to be turned down by the "big H," but, in retrospect, MIT, where I did get in, may have been a better fit.

Freshman year of college was spent trying to pick a major. I had little interest in engineering, but the sciences or possibly foreign languages and linguistics seemed like real prospects. What turned the tide was a few upperclassmen in my dormitory, who talked about the excitement of something called molecular biology. Never having had a biology course in my life, I had always considered biology to be a backwater of boring classification of plants and animals. When I overheard my elder housemates discussing their biology classes, it sounded like a new language. I was sufficiently intrigued to ask them about this emerging field, and pleasantly surprised to learn that the laws of chemistry and physics could be happily married to biological problems. So I decided to major in this wave of the future, but also took a lot of chemistry and physics classes to cover my bases.

During junior year, I wrestled with the question of what I wished to do with my life. For a biology major, this frequently amounted to a choice between medical school and graduate school. I had a hard time visualizing myself as a medical doctor—tethered to patients, clinics, and red tape. On the other hand, Wednesday golf seemed like a nice perk. In the end, the independence of a research career better suited me, and I opted for graduate school. In retrospect, I was somewhat naïve believing that medical school could not also lead to a successful career in research. But the decision was made and it was off to graduate school to carry out thesis work on bacterial genetics with Jonathan Beckwith at Harvard Medical School.

Jon's lab was extremely productive scientifically but was also deeply steeped in radical politics. Jon was an active member of Science for the People, a group that agitated for full oversight of all scientific enterprises by citizen groups. I was a detached spectator of the politics but was deeply drawn into the science of the lab. One of the most important lessons I learned there was that once you commit to studying a scientific problem, you have to be willing to use any available tool to further the project. Someone specializing in genetics may well reach a point in the project where a biochemical experiment is what is needed. The idea of avoiding an experiment because it is not in your bag of tricks is a fear that must be overcome for success in research.

It was in this spirit that I actually raised problems for the lab, and for myself. The tools of genetic engineering were becoming available during my years at Harvard. Restriction enzymes, tools that can snip DNA into precise bits, were staples of this new technology. I wanted to use these enzymes in my project, but certain lab members objected strenuously, claiming this would somehow lead us down a slippery slope toward moral depravity. To his credit, Jon assured me that I could go ahead, and I broke the ice of using tools of genetic engineering in the Beckwith lab.

Toward the end of my Ph.D. work, I had to decide on a lab in which to serve as a postdoctoral fellow. The postdoctoral period is an apprenticeship after the completion of the Ph.D. and before ascension to a faculty position, and it has become a sine qua non for advancement in biomedical research. I had become interested in higher organisms than bacteria, and phoned a well-known yeast researcher about the possibility of working in his lab. Disappointingly, he told me to spend some time in the library before troubling him again. I chalked up this curt response to likely political friction between him and my mentor. Next, I contacted Mark Ptashne at the Harvard Biological Laboratories. He was an innovator in the revolution in genetic engineering then under way but had little experience in yeast. He was enthusiastic about taking me, and his solution to my ignorance was that I could attend the next international meeting on yeast genetics and molecular biology. This fit nicely with the can-do attitude learned in graduate school, and I committed to his lab.

It may be a cliché, but I can only describe my experience in those postdoctoral years as that of a kid in the candy store. The new methods of genetic engineering opened possibilities to study yeast and even higher organisms at a genetic level never before possible. I tackled the yeast system with enthusiasm and began a study of how yeast genes were regulated. Even though Mark's lab had no prior experience with yeast, he and my lab mates were very supportive and interested in the work. Scientists often study one organism their whole career, and this is certainly true of many who study yeast. However, the strong affinity that I felt was not so much for the organism as for the scientific question being addressed. This attitude would prove useful much later when it came time to extend our studies of aging from yeast to other organisms.

First, it was essential to learn more about yeast aging. One of the questions Brian and I mulled over on the golf course was how we could isolate the relevant gene from his long-lived mutant strain. The surprising fact that this mutant was sterile meant that one and the same mutation caused the sterility and the longevity. This provided us with a trick to isolate this gene—we would identify a DNA fragment from a normal strain that would restore fertility when added to the long-lived mutant. Thus, Brian proceeded to isolate this piece of DNA, which, again, was one and the same as the longevity-conferring gene. Remarkably, it was a known yeast gene. Researchers stand on the shoulders of their predecessors, and over the past 20 years, yeast biologists had built up a collection of many mutants that were sterile. One of these mutants was termed SIR4, where SIR stands for silent *information regulator*. It turned out that the gene Brian identified was SIR4. A mutant form of the SIR4 gene was responsible for the sterility and longevity of the mutant strain.

Before considering what the SIR4 gene does, a bit of background is in order. Genes are the structural units of the DNA that determine the form and function of all organisms. The yeast genome contains about 6,000 genes, whereas the human genome has around 35,000. Most genes owe their effects to the fact that they encode proteins, which are polymers containing the 20 amino acids strung together like the multicolored beads of some cheap necklaces. The different sequences of the amino acids give proteins their distinct functions in cells. Since each position in a protein can be occupied by any of the 20 amino acids, the number of possible sequences of a typical protein of 200 amino acids is dizzyingly large, approximately 10 followed by 260 zeros! All told, the 6,000 yeast proteins represent only a tiny fraction of possible sequences.

Nature has evolved a genetic code that specifies the sequence of the amino acid residues of proteins from the sequence of nucleotides (A, G, C, and T) of genes. Cells decode the linear sequence of nucleotides within a gene three by three, where each triplet specifies the next amino acid to be laid down in the protein. Specific triplets at the start (ATG) and end (TAG or TAA or TGA) of genes mark the first and last residue of the protein. Not all genes are actively in the business of making proteins at any one time. Cells decide which genes should make proteins (i.e., "be expressed") by "Xeroxing" the DNA of the gene to create a molecule of RNA bearing the same sequence. It is this so-called messenger RNA that is directly used as the blueprint by the machinery that manufactures the proteins. Cells can then decide which genes are expressed, i.e., copied into RNA that is decoded to proteins. For example, red blood cells robustly express genes encoding hemoglobin, whereas skin cells express genes specifying keratin, and do not express genes encoding hemoglobin.

So the SIR4 gene codes for a protein that functions in yeast cells as part of a threesome with the protein products of two other genes, SIR2 and SIR3. The three SIR proteins are completely unrelated by their sequence, but they meld their talents in the troika to "silence" the expression, i.e., the copying of DNA into RNA, of a few genetic regions in yeast chromosomes, as described below. Our DNA is organized into 46 chromosomes, but yeast contains only 16. In normal yeast cells, the distinct genetic regions where the SIR proteins are attached are turned off, i.e., silenced, but are turned on when any one of these three SIR genes has been inactivated.

One of these specially silenced regions of the genome contains extra copies of genes that specify the two genders in yeast. Because the SIR troika normally silences these extra mating genes, the genes do not influence the mating behavior of cells. They are invisible. But in the SIR mutants, these extra mating genes become inappropriately expressed and cells adopt *both* genders at once. The poor, confused cells lose their sexual identity and their ability to mate. Thus, SIR mutants are sterile.

Yet another region where the SIR trio was known to exert its effect is at the ends of yeast chromosomes. The regions at the ends of chromosomes in organisms ranging from yeast to humans are termed telomeres. They are special DNA sequences that prevent erosion of the ends of chromosomes and also ensure that chromosomes are properly replicated. We were interested in telomeres because of reports that they become shorter in certain human tissues as we age, and this shortening was proposed as the cause of aging by Calvin Harley of the Geron Corporation in Menlo Park, California. The idea was that the erosion of the chromosomal ends would eventually strip away essential genes and interfere with chromosomal replication. We were initially excited that the SIR4-42 mutant might link telomere shortening to aging in yeast.

In the spirit of thoroughness, Brian checked to see whether the previously identified mutations in SIR4, i.e., those sterile mutants isolated by other labs, also extended the life span. This simple experiment gave a profound result: The other mutations did not extend the life span. We concluded that there must be something special about Brian's mutation, now called SIR4-42 because it was the 42nd strain in Nick's collection of mutants that was analyzed. The SIR4-42 protein must be changed in some subtle way, rather than being completely inactivated.

This line of reasoning also gave an additional insight. We knew that completely eliminating the SIR4 protein from cells, which is what the SIR4 mutants isolated by other labs do, causes sterility but does not make the cell live longer. Hence, the longevity conferred by SIR4-42 must have nothing to do with the mating genes or, for that matter, the telomeres, which are affected in the earlier mutants and SIR4-42 alike. For all their appeal, telomeres suddenly dropped off our radar screen, at least for yeast aging.

During this phase of the research, we moved the lab to the new biology building on the MIT campus. I was worried that the move would slow our research. In fact, we were back doing experiments within a week. The move also buoyed our spirits, because the new building was palatial compared to the old one, which a colleague of mine, Chip Quinn, memorably referred to as "eight floors of basement." Our new lab was at one end of the second floor with five rooms of lab benches along the outer walls. Scientists in the lab each had their own lab bench and attached desk. The rooms also contained a sink and a "safety shower" for the unlikely event of an emergency. The area around the stairwells was open and encased in glass walls, creating lines of sight and communication between scientists on different floors. This open area was flanked by common rooms for eating, seminars, or just hanging out. The spacious and aesthetically appealing nature of the new building was unusual for MIT. The MIT central planners try to make the buildings narrow, since they can then be more easily retrofitted for other purposes if departments decide to move again in the future. The architects and the building committee of MIT biology department faculty, however, fought for a wide building so that core facilities, such as cold rooms for storing cells, warm rooms for growing cells, and hot labs for radioactive work, could be placed in the center. Fortunately, the planners lost, and the big winners were the residents.

Once we were set up in the new building, we thought long and hard about Brian's mutant strain. A bold hypothesis sprang from our conclusion that the SIR4-42 mutation gave rise to a partially active protein, unlike the other mutations in SIR4, which completely destroyed it. We proposed that the SIR4-42 mutation inactivated the anchor that tethered the SIR2/3/4 troika to the mating and telomere loci, which is why silencing was abolished at those sites in the mutant strain.

However, we also figured that the SIR troika might normally function at another, as yet unidentified, region in the yeast genome, which we termed the AGE locus. We reasoned that the SIR4-42 mutant protein would retain the ability to join its SIR partners in the troika, and that the trio would continue to find the AGE locus. In fact, there would be more of the SIR proteins at the AGE locus than normal; the usual amount plus the stuff redirected from the telomeres and mating loci. The presence of the extra SIRs would increase silencing of gene expression at AGE. Since an increase in silencing extended the life span, we proposed that it was the expression of the AGE locus that caused aging in normal mother cells.

With excitement tinged by some trepidation, we prepared a manuscript on these novel findings in 1995, and submitted it to the prestigious journal, *Cell.* To our delight, Brian's paper was accepted for publication. Publishing in the "high profile" journals like *Cell, Nature,* and *Science* is fiercely competitive, and stories about the jockeying that goes on for space in these publications could fill another book. Our findings were well received by colleagues, in part because they fit one of the many schools of thought on aging; i.e., aging is caused by changes in gene expression as the organism grows older, in this case, the AGE locus.

CHAPTER 4

·····································

Identification of the AGE Locus

OUR YEARS HAD PASSED BETWEEN the time Brian and Nick started their experiments and the publication of the *Cell* paper. This slow start was an inevitable consequence of building something from first principles, but was nevertheless frustrating. Activities outside the lab kept me sane during this period, the most important of which was being a parent. It had been a delight watching my infant son, Jef, develop cognitive and then language skills. His given name was actually Jeffrey, and his cooler shortening of the name to Jeff and then Jef did not come till much later. I remember before he could talk, I stumbled upon the fact that he could identify any object in the house that I would name. Later, when he could talk, the honesty and clarity of thought of his young mind fascinated me.

My weekly schedule during these years was split into two. From Tuesday afternoon until Saturday, Jef was at his Mom's, and for the balance of the week he was with me. This meant all traveling and most work-related matters had to be scheduled during the Tuesday to Saturday block. This plan had two benefits. It forced me to be extremely organized and efficient in the days I did not have Jef, plus I could be a parent every week. That was a joy.

The *Cell* paper was the first widespread announcement that we were studying aging in yeast. But our theory of an AGE locus rested on a precarious foundation of data, and we worried whether it might be a flight of fancy. The only way to substantiate it was to prove the existence of the AGE locus. How could we identify this genomic site? Perhaps we could peer into cells using stains specific for the SIR troika and literally see for ourselves whether its location was altered in the SIR4-42 mutant. The lab of Susan Gasser in Lausanne, Switzerland, was the best in the world at visualizing the location of these proteins in yeast cells. Her lab could view bundles of the SIR troika stationed at yeast telomeres inside the nucleus, the part of the cell that houses the chromosomes.

The SIR proteins could be stained using homing devices called antibodies that selectively recognized them, among the thousands of proteins in yeast cells. Antibodies are proteins that we and other animal species produce to defend against intruders, like pathogenic viruses and bacteria. They have more recently become indispensable for the modern experimental biologist. In our case, an antibody that recognized and bound to SIR4 specifically was the critical experimental tool. This antibody was obtained from rabbits that had been injected with the purified SIR4 protein. The rabbits, quite sensibly, recognized the protein interloper as foreign and made antibodies that bound to it and it alone.

The location of the SIR4 protein in yeast was then determined by adding these anti-SIR4 antibodies to cells that had been attached to a glass slide. The antibodies found their target—something like a penguin locating its partner among thousands of apparently identical birds upon returning home for mating season. In order to see where the SIR4 antibodies were in cells, we added a second antibody, this one produced in goats, that binds specifically to the first. The goats are readily induced to make their own antibodies against the injected rabbit antibodies, because they see the invaders as foreign. The second antibody, the one from goats, is special because it is chemically attached to a molecule that emits fluorescence. It finds its target in yeast cells, the rabbit antibodies clinging to their SIR4 prey, and glows. The position of SIR4 thus lights up as a bright spot within yeast cells under the microscope.

Brian headed to Switzerland accompanied by his yeast strains to see where the troika resided in the SIR4-42 mutant. Gratifyingly, he quickly found that the troika had indeed changed residence in the SIR4-42 strain. Here serendipity again entered the scene, because the new location was easily recognizable as a site within cells—identified and well known for over a hundred years—called the nucleolus.

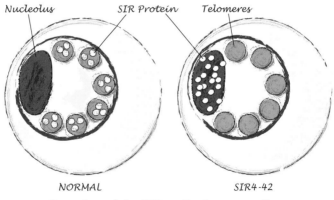

Location of the SIR troika in yeast cells.

The nucleolus is the site within the nucleus of cells that houses repeated copies of the same few genes, in fact, the most highly repeated DNA in the yeast genome. These genes, termed the ribosomal DNA (rDNA), are repeated 100–200 times in tandem and encode the two RNA molecules of the ribosome, a complex bundle of RNA and proteins present in many copies in cells. These RNA/protein balls are the machines that string together the amino acids in sequence to build proteins. They blindly follow the code specified by the nucleotide triplets within the RNA copy of genes. Highly repeated DNA sequences like the rDNA are intrinsically unstable, as discussed in Chapter 5: Chromosomes tend to lose some of the copies. The yeast cell is then faced with the daunting challenge of maintaining the stability of the rDNA, a challenge underscored by the fact that they span some 1–2 million base pairs of DNA, almost the entirety of the largest yeast chromosome.

The finding that the AGE locus was the unstable rDNA edged us toward the idea that it was the deterioration of this locus which somehow contributed to aging. This line of thought brought us back to the notion that aging is due to the decay of processes over time. By this view, the rDNA was the Achilles' heel in yeast, since its deterioration was the limiting factor in the life span. More generally, most current theories focus on the degeneration of one process or other as a primary cause of aging. One especially popular theory of aging proposes that all of these degenerative processes are due to oxidative damage over time. This view is so prevalent that I make a brief digression to discuss it before returning to our studies.

In the 1950s, the chemist Denham Harmon proposed that oxygen, which of course gives and sustains life in the brightness of youth, comes back and bites in the backside in the twilight years. Oxygen is normally converted into water as it is consumed during respiration, thereby creating the chemical source of energy for cells, called ATP (adenosine triphosphate). Harmon proposed that oxygen is occasionally hijacked during respiration to generate toxic by-products instead of water. These so-called oxygen radicals can react with the important macromolecules of cells, i.e., DNA, RNA, protein, and lipids, to trigger cellular damage leading to aging in the organism. There is now reasonable evidence that in at least some organisms, oxidative damage may indeed be one contributor to progressive damage and decline.

Age-associated damage is most severe in the cellular compartment in which most of the oxygen radicals are generated, the mitochondria. These membrane-bound structures are the power plants of cells and generate most of the ATP used for energy. Mitochondria bear features similar to bacterial cells and, indeed, are thought to have arisen because some primordial precursor of our cells swallowed a bacterium whole. These structures have their own tiny genomes, which is a vestige of the chromosome of their bacterial ancestor. Doug Wallace at the University of California, Irvine, is an expert on mitochondria and has championed the view that degeneration of this compartment due to oxidative damage is a cause of aging. He and his colleagues have demonstrated that damage to mitochondrial DNA and proteins really does accumulate with age. In fact, they were even able to make roundworms live longer by feeding them a synthetic compound that neutralized and detoxified oxygen radicals.

However, there is good reason to wonder whether oxidative damage is the be-all and end-all of aging. First, cells have active defenses against the toxic oxygen by-products in the form of enzymes such as superoxide dismutase and catalase. These enzymes are proteins that eat up reactive oxygen radicals before they can do harm to the various parts of the cell. There are also different cellular enzymes that repair the oxidatively damaged macromolecules themselves. Of course, it is possible that these resources for detoxification and repair are just not up to the job over the course of an organism's lifetime. But it is also likely that besides oxidative damage, there are other, unwanted saboteurs of cellular health, including radiation, heat stress, and a general tendency toward disorderly events, like the unfolding of proteins. What I take away from this area of study is that bad stuff clearly happens, but there are likely many different causes of aging in different organisms.

In our own studies, what seemed almost magical was that yeast genetics led us to something that promoted *survival*. The SIR troika works to *counteract* aging. So we pondered the question of which was more likely to provide insight into the aging of higher organisms—the deterioration of the rDNA as a cause of aging or the protective activity of the SIR troika to promote survival. It did seem strange to imagine that the SIR trio would promote survival in an organism in which aging might not be caused by rDNA decay. Was it possible that the SIRs somehow had a broader activity in nature than the protection of the rDNA? To even consider this question, we would first have to figure out the exact nature of what goes wrong in the rDNA of aging mother cells, and then how the SIR troika protects against this.

CHAPTER 5

The Rise and Fall of rDNA Circles and the Emergence of SIR2

W E FOCUSED LASER-LIKE ATTENTION on what might be going wrong in the yeast rDNA with age. The reason the repeated rDNA sequences are unstable is that they are in danger of being lost from the chromosome by a process termed recombination. Recombination is the breaking and joining of two chromosomes at comparable positions along their length to generate hybrid chromosomes. These hybrids create genetic diversity when passed on to the emerging population of the next generation.

Recombination takes place between stretches of DNA with the same sequences. Typically this is only the case at comparable positions on the two chromosomes. If, however, a certain stretch of DNA is repeated at two or more locations on the same chromosomes, recombination can take place between them. This leads to the loss of that DNA which falls between the repeated stretches on that chromosome.

The rDNA is an exceptional chromosomal region in this regard because the same gene is repeated many times in a row on the same chromosome. The organism does not wish to carry out unrestrained homologous recombination in this region, because gene copies could be lost in the shuffle. To stay in business, a cell must retain the right number of copies of the rDNA; too low a number is bad news because cells are unable to produce enough ribosomes. There are mechanisms to repress recombination of the repeated rDNA array. In fact, we knew from the earlier work of Shelley Esposito's lab at the University of Chicago that yeast cells, in particular, had a defense mechanism to thwart this tendency of cells to shuffle the rDNA sequences by recombination. Remarkably, the central player in this defense was none other than one of the three SIR genes. SIR2, but not SIR3 or SIR4, repressed recombination. In SIR2 mutant strains, the rDNA copies became much less stable and were jettisoned more frequently by recombination. Thus, SIR2 possesses an additional function, on top of silencing at mating genes and telomeres. As shown below, this extra function foreshadowed that SIR2, and not the other SIR genes, would turn out to be the most important gene of all in dictating life span.

The idea that aging, in general, might be due to the gradual loss of rDNA copies had actually been proposed decades before by Bernard Strehler, a long-time researcher on aging from the University of Southern California, from his studies of the rDNA in aging dogs. One of the quaint things about the aging field is that so many models have been proposed over the years that it is likely that any new idea, however baroque, has already been put forward. There was one way in which Strehler's model did not apply to yeast. Daughter cells that bud off mothers are rejuvenated; i.e., they will live a normal life span. This is essential, of course, because otherwise the life span of yeast would have spiraled down to zero long ago and the species would be extinct. The Strehler model cannot explain this rejuvenation since, according to that model, daughters of aging moms should also inherit the deficient rDNA array and should have their own life spans reduced accordingly.

At this point in the project, it was evident that we might really be on to something, and that the lab's investment in aging needed to expand. Giving seminars about research in the lab is one way of drumming up new recruits. During a seminar visit in Australia, where I talked about our new work on aging, I met a student at the University of Sydney and convinced him to come and work in my lab as a postdoctoral fellow. This particular fellow, David Sinclair, had completed a very productive Ph.D. thesis in Australia, and he came to MIT bubbling over with energy and ideas. He bounded into the lab on his first day as though out of a Crocodile Dundee movie and greeted each of us with a friendly "hello mate." A while after his arrival, he prominently displayed a news clipping from back home, which showed a young child at the zoo watching two kangaroos conspicuously copulating while his surprised parents looked on. Not all the folks in the lab were so amused, and the photo quickly came down.

In some experiments, David joined forces with Kevin Mills, who, with another talented but not fully mature student, David Lombard, helped earn our lab the reputation as the biology department frat house. Kevin and David both held Beavis and Butthead as cultural icons. The team of Sinclair and Mills thrived because each brought out the best in the other. Kevin's friendly barbs spurred David to generate high-quality data rapidly. And the favor was returned in kind.

In late 1996, postdoc Brad Johnson and I were struggling with the problem of trying to apply Strehler's model when David bolted into my office in the grip of inspiration. He thought that aging mother cells, instead of losing copies of rDNA as Strehler had suggested, might actually be *gaining* them. How could this be possible? He noted that each of the 100–200 rDNA repeat units contained a DNA site called an origin of replication, which could facilitate the initiation of DNA replication every cell division. Therefore, if two regions of rDNA from the *same* chromosome underwent recombination, and instead of being cut out and thrown away, simply popped out of the chromosome, a DNA circle would be created. This circle would arise because the two ends of the DNA segment excised from the chromosome are joined together by the recombination process. Moreover, this circle would have the ability to replicate outside of its normal chromosomal abode. So at every cell division, when the chromosomes duplicate, the circle would also duplicate.

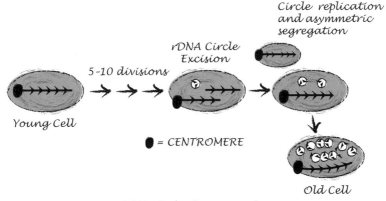

rDNA circles in yeast aging.

How would this cause a significant increase in rDNA copies? Nature ensures that the bona fide chromosomes, once duplicated, separate faithfully between the mother and daughter; i.e., one chromosome to each cell. A special chromosomal region called a centromere ensures this separation. But the centromere of the yeast chromosome XII happens to lie outside the rDNA repeats. Since the rDNA circle would lack a centromere, perhaps both of the duplicated copies would disregard the normal rules of partitioning. If rDNA circles happened to remain solely within the mother cell after cell division, the number of these circles would double each time the mother replicated its DNA and produced another daughter. Once the first circle popped out of the chromosome, circles would build up in a precisely timed manner. After 10 divisions, the number of circles could approach 1000.

How might this buildup cause aging? The excess of rDNA could be toxic for a number of reasons; perhaps most simply it would create an excess of ribosomal RNA. This would upset the delicate balance between the RNA and protein components of the ribosome, which are encoded elsewhere in the genome. Most importantly, David's model explained why the daughters are rejuvenated: They fail to inherit any of these circles.

In order to do an experiment to test the model, David needed to isolate old yeast mother cells in a pure form. This task was yet another little obstacle that stood in the way of successfully studying what was different about old yeast cells compared to young cells. In a test tube of growing yeast cells, the rules of statistics dictate that half of the cells will have divided zero times (i.e., are recently formed daughters), one-quarter have divided one time, one-eighth two times, one-sixteenth three times, etc. If you continue this mathematical progression to include cells that have divided 20 times, i.e., are truly old, you find their frequency is but one in a million. These are the cells we wished to study in pure form. How could we separate them from their younger brethren?

Another postdoc in the lab, Tod Smeal, helped develop a method that we used to isolate these old cells. Tod is the son of Eleanor Smeal, former president of the National Organization for Women and founder of the Feminist Majority. Like his mother, Tod is not bashful about telling you what is right and wrong in the world of politics and government. Unlike his mother, Tod's purview extends to science. The method Tod developed depends on the fact that when yeast cells bud, the outer surface of the daughters consists entirely of newly made material. The mother, meanwhile, retains its old shell. Thus, if one chemically tags the outside of cells with a molecule called biotin, that label will stay entirely with the mother cell after a round of cell division. Their daughters, and all subsequent progeny of these daughters, are biotin-free.

Accordingly, Tod labeled a culture of roughly a billion yeast cells with biotin. If the culture were allowed to grow for enough time to allow, say, 20 generations of cell division, the only cells that would contain the biotin label would be those mothers that were tagged in the first place, but now they would be 20 generations old. These "old" cells could now be separated from the rest by mixing in magnetic beads that were coupled with a kind of glue called avidin that sticks very tightly to biotin. Now that very advanced instrument called a magnet would pull out these old cells.

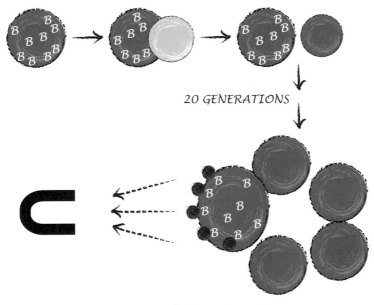

20 GENERATIONS

Isolation of old mother cells.

With the old cells in hand, David could extract their DNA and place it in a semi-solid gelatin-like matrix in an electric current. This technique is called gel electrophoresis and was used to search for the presence of any of these rDNA circles. The electric current moves the negatively charged DNA molecules through the gel, and different DNA molecules are separated by their size, with the smallest DNA molecules moving the fastest. If they were present, rDNA circles would migrate to a characteristic location in the gel based on their size. When I was a young postdoc at Harvard, I ran two of these gels a day, week in and week out, to analyze DNA. There is one overriding rule to keep in mind: Do not touch the gel after you turn on the current. I have heard of fatalities when this rule has been breached.

In a rapid series of elegant experiments, David found that these circles were totally absent from young cells, but became quite abundant in old cells. Moreover, if he employed a genetic trick to increase the rate of formation of these rDNA circles in live yeast cells, the life span of mother cells was shortened. Thus, rDNA circles really did appear to be a cause of aging. There was a lot of excitement about these findings when they were published. They suggested a molecular mechanism of aging. Even my chairman was warming up to the idea of studying aging and told me that the rDNA circles model was "the first notch in your belt."

Some researchers think we may have overemphasized the importance of rDNA circles in yeast aging and excluded other possible mechanisms. I agree to a point. As mentioned above, an inappropriate level of expression of the rDNA genes may itself be the critical determinant of aging. The circles may only indirectly contribute to aging, because they promote the expression of too much ribosomal RNA. It is also possible that still other mechanisms completely unrelated to the rDNA are critical, although I have not seen any compelling evidence to date.

One of the most unusual aspects of working on aging is that there is considerable attention from the media. Our society seems to gobble up news of any new treatments, medical or cosmetic, purporting to slow the effects of aging. The media cater to this audience, even covering serious matters, like advances at the forefront of aging research. As a result, our lab members witness occasionally film crews preparing a piece for public TV or another outlet.

Members of the print media also phone, and they reported on our paper on rDNA circles on the Friday it was published. I was even asked to appear on "Good Morning America." The down side was that I had to turn up at a local TV studio at 8 a.m. Sunday for the live satellite hookup. Nonetheless, it was fun being interviewed by Willow Bay, the anchor, 200 miles away in New York while her celebrity guest, Eartha Kitt, sat next to her in the studio. As I dressed for the interview, a wave of impishness broke over me and I chose to wear my *NY Times* gift necktie, a Christmas present to subscribers. The tie, called Absolute Dad, was festooned with vodka bottles designed to resemble sperm, complete with their wriggling tails. When I returned home, my parents called and said they enjoyed the broadcast. They didn't mention the tie.

A few weeks later, the rDNA circle model of aging appeared on the front page of the *New York Times* science section accompanied by a long article by Nicholas Wade, the science reporter for the *Times*. I had gotten to know him from phone conversations in which he asked me to comment on other stories he was preparing on aging. As often happens in life, the period of basking in the glow of fame was very short, literally approaching the 15 minutes promised by Andy Warhol. The same day my article came out in the *Times*, a biotech company working on aging issued a bulletin announcing a major breakthrough. The Geron Corporation, and their collaborators at the University of Texas Southwestern Medical Center, Dallas, had figured out what caused the Hayflick limit in cultured human cells, i.e., the inability of these cells to divide indefinitely. It was the shortening of telomeres to a length that meant the death of the cells. If the researchers supplied these cells with the enzyme that maintains the telomeres, an enzyme called telomerase, the cells divided endlessly.

Within hours of the appearance of the article on rDNA circles, Nicholas Wade was again on the phone, not to discuss our circles, but rather to discuss the importance of Geron's new findings on telomeres. All of the news broadcasts that afternoon and evening focused on telomeres, and our own little story was already forgotten. The public's fascination with the telomere model of aging is understandable, because the model provides a timer for cell division and is simple to comprehend. Although our own experiments argued against a role of telomere shortening in yeast aging, they did not speak to a possible role for telomeres in mammalian aging. So telomere shortening could still cause human aging. But, in my view, cultured cells constitute a rather unnatural system. In the whole animal, these cells would be sloughed off before their telomeres were too short and they would be replenished by the differentiation of precursor cells that *do* have telomerase. The cultured cells are cut off from this source of renewal and left hanging in the breeze.

There are other reasons to think that telomere shortening may not play a critical role in the aging of animals. First of all, telomeres are not short at all in several strains of mice that have been tested. The telomeres can be made shorter, certainly, by deleting the gene for the telomerase enzyme in mice. The telomeres get shorter in these genetically altered animals in an interesting way. Because of the mutants' inability to maintain the telomeres in the germ line, i.e., the egg cells and sperm, the ends of the chromosomes are whittled down progressively in successive generations as these animals are mated to each other. Eventually, by the fifth generation or so, telomeres in many organs became critically short and the tissues with dividing cells degenerate. So, yes, if animals are manipulated in the lab to make the telomeres become short, the animal will get sick. But these sick animals do not duplicate normal aging. Most importantly, the evidence is still lacking that telomeres ever reach this critically short length in normal rodents, i.e., with the telomerase enzyme intact. This is also true in humans, where the telomeres do shorten a little with age, but may never become short enough to cause harm.

However, at least one interested party read our article about rDNA circles. I received a phone call from someone who wanted to put me on notice that he had already patented the claim that "DNA causes aging." As a token of good will, he offered to send us a sample of a skin cream he had developed. The active ingredient was the enzyme that chews up DNA called DNase. I tried not to betray my reaction that the idea of degrading DNA indiscriminately was ludicrous. We were then in the Unabomber era: When the package arrived, no one in the lab was encouraged to go near it and it was never opened.

Flush with our rDNA mechanism of aging, Brad and I were quite keen to begin to look for rDNA circles in other organisms. But the yeast model seemed to depend on the peculiar, asymmetric cell division by budding only found in that organism. We wondered where to find an analogous situation in animals, which are made up of some organs with symmetrically dividing cells (skin, blood, intestine, germ line, etc.) and other organs with cells that do not divide at all (brain, heart, skeletal muscle). One possible analog of the asymmetric cell division seen in yeast might be stem cells, precursor cells that divide to provide the continuous supply of differentiated cells that make up certain specialized tissues. Stem cells are difficult to isolate, however, not least because they lurk in deep recesses of organs.

Our experiments involved grinding up whole organs (in this case from rodents), including the stem cells, and applying very sensitive assays to detect the appearance of any rDNA circles with age. In a somewhat ghoulish flourish, we also examined a limited number of human tissues obtained at autopsy. In this experiment, I was out of my element and we were able to obtain these tissue samples only because Brad had an M.D. degree, in addition to his Ph.D. His contacts in the world of medicine paid dividends. After two years of study, we had found, at best, low levels of these circles and certainly no trend toward their accumulation with age. We also observed that no other kinds of DNA circles accumulated.

Not only were rDNA sequences not gained, but they were not lost, with age either, as would be predicted by the Strehler model of aging. I began to feel rather guilty for dragging Brad through the morass of these negative experiments. Fortunately, he displayed his keen acumen by ignoring my pleas for more experimental data and starting his own parallel, more productive, lines of research.

Most importantly, our studies on rDNA told us that there was no universal mechanism that caused aging. The rDNA gymnastics that are central to yeast aging do not occur in aging animals. Depressingly, aging appeared to be more haphazard than I had thought, just as the evolutionary biologists warned us. On the other hand, Cynthia Kenyon at UCSF had by now independently followed up the long-lived age-1 mutant of Klass and Johnson with the clearest demonstration to date that a single gene mutation could greatly extend the life span of worms. We return to these worm genes in Chapter 7.

Were we sinking in a swamp of many genes and many causes of aging or not? Well, I did worry that all might be lost, but the next phase in the research on the SIR genes and aging would lead us to the conclusion that there may indeed be a simple, universal mechanism. The key concept harks back to the idea that nature did not create this mechanism to *cause aging*, but, rather, to slow aging in times of scarcity to *promote survival*. Some simple experiments in yeast turned the corner for us at this critical juncture.

Two newer graduate students in the lab, Matt Kaeberlein and Mitch McVey, were in the process of taking our genetic studies in yeast to a higher level of understanding. Matt came to the lab brimming with self-confidence, even by MIT standards. And why not—he was smart and displayed mature judgment beyond his years. I would often pop out of my office to get his take on a new paper appearing in a scientific journal. His work ethic caught my attention early on. We would discuss future life span experiments on a Friday, and he would have the results on Monday. The only problem was that I usually did not see him in the lab on weekends, raising serious questions about his experimental technique.

Just as I was getting really concerned, Matt revealed he had "borrowed" a microscope from the lab to set up his own lab at the marine biology field station where he and his wife lived. He could thus work all weekend on life span assays without ever leaving home. Mitch was as upbeat, balanced, and likable a guy as I have ever had in the lab. And he loved to teach. When I taught genetics class to MIT undergraduates, Mitch was one of about ten teaching assistants who ran their own study groups to work through sample problems to practice for exams. Students were free to attend any study group they pleased. By the end of the term, Mitch had a group many times larger than the others, which we affectionately called Mitch's cult.

Matt and Mitch worked closely together to study the individual roles of each of the SIR genes, SIR2, SIR3, and SIR4, in yeast aging. By mutating these genes to inactivity, they found that SIR2 was of unique importance: Its removal shortened life span drastically, whereas removal of SIR3 or 4 exerted at best minor effects. Moreover, adding just one extra copy of the SIR2 gene to normal yeast cells gave a substantial increase in life span. This striking finding meant that the amount of SIR2 in normal cells is what limited their life span. We could now easily understand how the SIR4-42 mutation extended life span—it brought more SIR2 (in this case as a part of the troika) to the rDNA. But what did SIR2 do at the rDNA? The simple answer harks back to the findings of Esposito at the University of Chicago showing that SIR2 represses recombination and stabilizes the rDNA repeats. This means that when the level of the SIR2 protein at the rDNA is high, cells have a lower tendency to form that first rDNA circle, which would initiate the aging process.

The importance of these genetic findings was that they narrowed our focus to SIR2 and SIR2 alone. This would lead us to a fundamental discovery on SIR2's mode of action, suggesting a mechanism by which nature may link survival to the availability of food.

CHAPTER 6

A Molecular Blueprint for Prolonging Life Span

WHAT WOULD TURN OUT TO BE the most important finding of all came about by accident. We knew that SIR2 mediated silencing of gene expression at the rDNA somehow, but it seemed important to delve more deeply into exactly how SIR2 performed this silencing act. We were like medieval time-travelers, who could easily see that automobiles were propelled by a pungent liquid obtained at gas stations; but learning what actually makes the car go would require lifting the hood and studying all of the moving parts.

There was already rapidly growing interest in SIR2 in the research community. The lab of Jef Boeke at Johns Hopkins Medical School found DNA sequences related to SIR2 in other organisms, and it was becoming clear that genes bearing very similar sequences to SIR2 were present everywhere—from bacteria to humans. There is no free-living organism I know of that has been shown to lack a SIR2-like gene. It was eye-opening that a gene involved in the seemingly baroque process of yeast silencing is universal. Whatever biochemical activity SIR2 possessed must have been preserved over more than two billion years of evolution. This implies that the activity must be of fundamental importance to living organisms. This conservation notably does not apply to SIR2's partners, SIR3 or SIR4. Any sequences related to these two genes are curiously lacking from all organisms except the most closely related budding yeasts; their activities are evidently less fundamental.

But how do genes get silenced? It is well known that genes do not exist as free DNA in cells, but are bound to a ball of proteins, called histones, around which the DNA is wound. This DNA–histone complex, called chromatin, wraps the DNA into a form compact enough to fit into a cell's nucleus. DNA from one of our own cells in its fully extended form would stretch across the length of a yardstick! Even in the compact form of chromatin, the DNA is generally accessible to proteins in nuclei; for example, the enzyme that copies DNA into RNA. However, in chromatin that has been silenced, this DNA–histone complex has been squeezed even further into such a tight structure that access to this enzyme has been blocked. Thus, gene expression is silenced. This inaccessible state of silenced chromatin not only reduces gene expression, but also lowers rates of recombination for the same reason; the enzymes that catalyze recombination cannot get to the DNA.

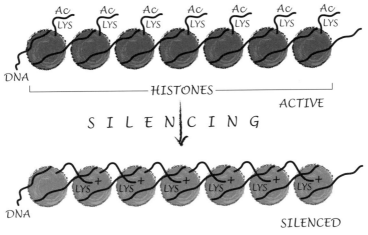

Active and silenced chromatin.

How does the chromatin get this extra squeeze? How exactly does it get compressed in this way? Two of the histone proteins called H3 and H4 have free ends called tails, which contain the amino acid lysine at a few key positions. In normal, expressed chromatin, these lysines have small molecules called acetyl groups attached to them, but in silenced chromatin these acetyl tags have been removed. These tags are the critical determinants for silencing, and their removal is what triggers the formation of the tight, inaccessible structure of silenced chromatin. This process can be likened to kicking out a doorstop and having a spring-powered door slam shut.

On the basis of this fact, Princeton's Jim Broach had the idea that the critical feature of SIR2 might be an enzymatic activity; i.e., the ability to remove the acetyl tags from the histones. A way to test this hypothesis would be to purify the SIR2 protein and place it in a test tube along with histones bearing the tags. If SIR2 were, in fact, a deacetylating enzyme, it would remove the tags from the histones in this setting. However, numerous labs, including Jim's, tried this experiment and observed no activity. Therefore, the whole world concluded that SIR2 wasn't such an enzyme.

A competing hypothesis came along. It proposed that, yes, SIR2 had an enzymatic activity, but it was very different from that previously suggested. Roy Frye at the University of Pittsburgh could demonstrate that SIR2 would carry out a reaction in a test tube that broke apart a molecule called NAD (nicotinamide adenine dinucleotide) and transferred half of it, called ADP-ribose, to another protein, which was also present in the test tube. Frye first pursued this hunch because an earlier study suggested that a bacterial cousin of SIR2 interacted with nicotinic acid, the other half of NAD. Frye's experiments were intriguing but not altogether persuasive. Although the demonstration that SIR2 functioned as an ADP-ribose transferring enzyme was clear, the enzyme functioned only feebly in his assay. Nonetheless, a postdoc in my lab, Shin Imai, and I were sufficiently piqued to try to follow up on Frye's findings in early 1999.

Shin had just arrived from Japan and was an unlikely blend of respectful obsequiousness and singular, in-your-face purposefulness. This latter trait became obvious when our building was evacuated due to a fire alarm and Shin continued his experiment on the sidewalk, accompanied by his ice bucket and test tubes. Shin's experimental data were always very neat and clear-cut. Scientists vary widely in their ability to generate clean data, much as teenagers vary in their tendency to maintain order in their rooms. In this case, Shin's position at the far end of this spectrum would be a boon for the discovery about to happen. He was aided in this study by another graduate student in the lab, Chris Armstrong, who prided himself on his encyclopedic range of general knowledge. He could hold court on the essence of successful cricket farming, for example. The fact that almost all of this knowledge was totally useless to us on this project was a source of further pride.

We thought that Frye might be essentially correct, but that his data were weak because his experimental system did not contain the correct protein to receive the ADP-ribose. Perhaps the histone tails were the normal protein recipients to which SIR2 added ADP-ribose. If this were true, it could mean that the ADP-ribosylation of histones in cells triggered silencing and not their deacetylation.

We devised an experiment to observe the transfer of ADP-ribose by SIR2 from NAD to short proteins corresponding to the histone tails. In fact, these short proteins weren't taken from histones themselves, but were synthesized in the lab and could be purchased from the manufacturer, Upstate Biotechnology. Their catalog even offered them in two flavors, with or without the acetyl tags. Shin asked me which version of these proteins to order and I said—without giving it much real thought—it should not matter. So he bought both, in what would turn out to be a most fortunate choice. Meanwhile, we purified SIR2 from bacterial cells that were engineered to manufacture abundant amounts of the yeast protein. In our first experiments we mixed the purified SIR2, the new, store-bought tails, and NAD containing a radioactive atom at a strategic location. By examining the reacted material in the gel, we could observe the transfer of the radioactivity from the low-molecular-weight NAD to the higher-molecular-weight histone protein. This weight gain would be evident because the radioactive ADP-ribose—now attached to the histone—would migrate slowly in the gel.

This highly sensitive method allowed us to detect very low levels of transfer of radioactivity to the histone. Like Frye, we observed the same feeble transfer of ADP-ribose. But there was one additional clue that no one had seen before. The acetyl-tagged histone tails received more ADP-ribose than the untagged tails, although still a relatively puny amount. This was not the big breakthrough we were hoping for, but this finding did seem like a small step forward: SIR2 could somehow tell the difference between tagged and untagged histones. The question of whether the ADP-ribose transfer activity was important still seemed very much up in the air.

Around this time, there was a frenzy of activity in other labs to pin down the relevant enzymatic activity of SIR2. At Harvard, Danesh Moazed, a superior biochemist, was convinced that Frye was right, and he wanted to publish a new study that purported to substantiate Frye's claims. Moazed's finding was that a mutation that changed a specific amino acid in SIR2 abolished the ADP-ribose transfer activity in the test tube and also clobbered the ability of a SIR2 gene bearing the same mutation to silence genes in yeast cells. Since his lab and ours were in contact, we thought this might be a good opportunity to submit an accompanying manuscript of

Shin's findings showing that SIR2 could discriminate between tagged and untagged histone proteins. However, I was very nervous about committing our findings to paper, since SIR2's activity was still anemic, and I had a gut feeling that there was an important twist in the story that we did not yet comprehend. Nonetheless, Shin's insistence and Danesh's desire not to hold off submitting his paper impelled us to prepare a manuscript.

About a week after we submitted the paper, we adopted a different strategy to analyze the product of the biochemical reaction catalyzed by SIR2. Although transfer of the radioactive label indicated that a chemical reaction took place, it did not prove that it was precisely ADP-ribose that was transferred to the protein. Maybe it was some other fragment of the NAD that happened to include the label. To prove that it was ADP-ribose that was transferred, we needed to come up with a more precise method to analyze exactly what we were getting from the reaction.

Mass spectrometry is valuable precisely because it can assign the exact molecular weight of anything tested. The material to be weighed is inserted into a special chamber in this machine, electrically charged, and made to fly through the air like a football kicked at the goalpost. The distance it flies indicates its molecular weight. The machine assigned the weight of substances tested in units called Daltons, named for the great 18th century English chemist. If ADP-ribose got added to our tagged protein, its molecular weight would change from 4740, the size of the tagged starting protein, to 5262, the size of this protein plus an ADP-ribose. In this experiment we didn't have to use radioactively labeled NAD, and so could use much larger amounts.

However, we would first need to separate the reaction products from the starting material, which could be done by a standard technique in biology, called chromatography. This method of sieving allows the separation of different proteins based on their different physical properties, but it does not yield precise information about their size. This new approach of analyzing the reaction products made us regular visitors at the "protein core" facility in the MIT Cancer Center, which houses the heavy equipment, i.e., the mass spectrometer and the chromatography apparatus. Entering that facility seemed a little like walking into a control room at Cape Canaveral, our familiar walls of plaster and paint replaced with a dismaying array of buttons, levers, and dials. The director of the core, Richard Cook, was very amiable and patient, which was just as well, because we needed some serious hand-holding in this new world of instrumentation.

We were excited to see by chromatography that the reaction products contained an extra something not found in the starting material, which we presumed was the ADP-ribosylated histone protein. We gave a sample of this now purified reaction product to Heather Amarosa, the lab technician who operated the mass spectrometer, as Shin and I sat by her side. Strangely, the machine gave a reading of 4698 instead of 5262. Shin felt we must have done something terribly wrong. I thought for a few minutes, then broke out in laughter. 4698 was the exact size of the starting protein (4740) *minus* an acetyl group (42). In other words, in our reaction, SIR2 was not adding ADP-ribose but was *removing* the acetyl tag. I laughed out loud because I knew that the paper we just submitted had completely missed the point—SIR2 was a deacetylase after all. I felt a little like Jed Clampett, rifle in hand, as he watched the crude oil bubble up from the underbrush. Serendipity, I love you!

Why other labs had failed to demonstrate this activity over the previous several years was immediately obvious. Our reactions contained not only SIR2 and the tagged protein substrate, but also NAD. This was because the activity we were looking for was not deacetylation at all, but ADP-ribose transfer. Perhaps NAD was somehow essential to trigger the deacetylase activity of SIR2. There would have been no reason whatsoever for the earlier researchers to have NAD present in their reactions. Of course, we still needed to prove our idea by repeating the experiment, but this time in the absence of NAD. This we did immediately, and when we analyzed this reaction by chromatography, the histone with the acetyls removed was nowhere to be found.

Wow, SIR2 was an NAD-dependent histone deacetylase! Such an enzyme had never been described before. The NAD was evidently the boot providing the heft to kick out the doorstop. A final question was why Moazed's mutation crippling his enzymatic activity also abolished silencing in yeast cells. Well, it turns out that this mutation, which knocked out the ADP-ribose transfer, also kills the histone deacetylase activity, explaining why the mutant protein cannot silence. The relevance of the NAD-dependent deacetylase to silencing in living cells was underscored by additional genetic experiments in our lab and that of Jef Boeke. I tried unsuccessfully to convince Shin that we should retract the ill-conceived paper we had submitted, and was for the first time gratified when we subsequently received savage reviews from the journal. Submitted manuscripts must suitably impress two or three of your peers, experts in the field, who are asked to review them. Otherwise the journal will turn them back. In this case, the peer reviews assured that the paper would not be published, thus burying the ill-begotten manuscript forever.

Equally interestingly, we were able to carry out very similar experiments with a closely related SIR2 gene that we cloned from mice, called SIRT1 (SIR Two number 1). By isolating this mouse gene, Shin was able to express the encoded mouse protein in bacteria, purify it, and test it in our assay. Strikingly, it possessed the same enzymatic activity as the yeast protein, that of an NAD-dependent histone deacetylase. This finding proved that the SIR2 sequences—present everywhere—really do encode this unusual enzymatic activity in a

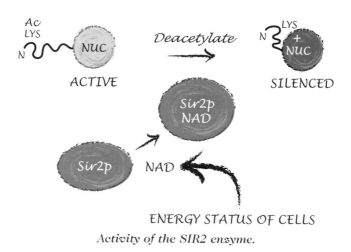

ENERGY STATUS OF CELLS

Activity of the SIR2 enzyme.

wide swath of nature's creatures. We were left to ponder the other, weak activity of SIR2, the ADP-ribose transfer, which got us started in the first place. It now seems but a curious side reaction, which may result from the fact that every time SIR2 removes an acetyl tag from a protein substrate, it splits a molecule of NAD. This splitting may rarely result in the inconsequential transfer of the ADP-ribose to a nearby protein.

Within a few days, the importance of this new activity for SIR2, NAD-dependent histone deacetylation, began to come into focus. NAD is a molecule that is present in all living cells and plays a critical role in many metabolic reactions. These metabolic reactions fall into two categories based on their chemical nature, oxidation and reduction. When molecules are oxidized, like the rusting of iron, electrons are removed from that molecule and donated to oxygen. In oxidation reactions involving metabolic intermediates in cells, electrons are donated instead to NAD to form NADH. In reduction reactions, conversely, electrons must be added back to the metabolic intermediates. NADH is the electron donor in these latter reactions, thereby regenerating NAD. Without these NAD/NADH conduits, metabolic reactions in cells and life itself would come to a screeching halt. All that matters from our point of view is that there is a link between the level of NAD and the metabolic condition of the cell.

By virtue of its enzymatic activity, that of an NAD-dependent histone deacetylase, perhaps SIR2 is able to sense the metabolic state of cells and at once connect metabolism to silencing and, more importantly, to aging. Everyone in the aging field knew of the long-standing relationship that had been established between metabolic activity and the rate of aging. Most strikingly, restricting the food intake of animals was known to extend their life span up to twofold. This fascinating subject of dietary or calorie restriction (CR) is the topic of chapter 10. Although CR had been first shown to extend life span in the 1930s, how it did this was a complete mystery. A possibility—riveting in its simplicity—was that SIR2 could be the regulator at the heart of this mechanism, by sensing the metabolic rate of an organism and fixing its pace of aging accordingly.

I was even more anxious than usual about getting these new results published. In early December, 1999, we submitted two papers to *Nature* on the work. The first paper dealt with the biochemical reaction, and the second compared the effects of SIR2 mutations on the enzymatic activity in a test tube to silencing in yeast cells. One month later we got back three reviews. Two were supportive, with the customary suggestion for changes, which included shortening the presentation to one paper. The third was negative, simply because the reviewer did not believe our claim that SIR2 had this unusual activity. We must be doing something wrong. I virtually had to go on my knees to assure the editor that our findings were correct.

It was published in February, 2000, and went down very well indeed. At our weekly department colloquium my chairman turned to me and said "nice paper" with a nodding smile. The journal could not give us the space to speculate fully on the implications of this enzymatic activity for aging. So, I immediately began to write a review to do just that. It was published in *Genes and Development* in April, 2000. There was some press coverage of the *Nature* paper, but the technical nature of the findings rendered them less media-friendly than simpler data about life spans. Nonetheless, I considered the work our finest moment: It spelled out a bio-chemical mechanism to set the pace of aging to the availability of nutrients.

Our experiments to this point told us there was no universal cause of aging. On the positive side, SIR2 had emerged from yeast genetics as a key determinant of life span, at least in that organism. The actual function of the SIR2 protein, an NAD-dependent histone deactylase, was indeed provocative—this protein could link life span to metabolism. Since this connection was widely observed in nature, we wondered whether SIR-related genes might determine life span in a wider range of the biological kingdom.

CHAPTER 7

SIR2 as a Universal Regulator of Survival

I WAS FEELING VERY SATISFIED with our SIR2 studies in yeast mother cells and in test tubes. This factor seemed to promote longevity and, quite possibly, also adjusted the rate of aging according to environmental factors, say how much food is available. The idea that rDNA was an organizing principle in aging had been dashed, and seeking to find any general causes of aging seemed like a futile venture.

So we were now ready to fully consider the transcendental question, not what causes aging, but what promotes survival. A survival factor could provide a simple mechanism to link the rate of the aging process to the environment, whatever the more immediate causes of aging might be. Is there such a factor? In yeast at least, we had a candidate, SIR2. Was it possible that SIR2 genes have been selected by nature to regulate aging in organisms other than yeast, even though the causes of aging might differ? I posed this question to a postdoc in the lab, Heidi Tissenbaum, several years ago, and she embarked on a series of experiments that would address whether SIR2 genes were also regulators of longevity in an animal species, the roundworm, *Caenorhabditis elegans* or *C. elegans*. This bold experiment was feasible because of the extraordinary strides taken in genetic studies of *C. elegans*, which were first developed by Sydney Brenner in Cambridge,

United Kingdom. The fact that Heidi had done her thesis work in the lab of Gary Ruvkun at Harvard, a leader in *C. elegans* aging research, also did not hurt the cause. The clincher was the generous offer of space and advice from one of the leading *C. elegans* labs in the world, that of Bob Horvitz, which happened to be right upstairs from us at MIT.

Heidi was an energy-charged young scientist. Most importantly, she was an unflinching optimist. On the face of it, the notion that SIR2 regulated aging in worms seemed preposterous, since we knew that SIR2 slowed the deterioration of the rDNA in yeast, but we saw no changes in the rDNA in aging worms. Even more fundamentally, the 959 somatic cells of the adult worm are all nondividing, which means that the two-week life span of adult worms is due to the decay of cells that do not undergo cell division at all. This pesky fact of biology seemed to indicate that the yeast model of replicative aging, i.e., aging caused by continuous cell division, was totally inappropriate for worms. On the other hand, the enzymatic activity of yeast SIR2 was just too pretty not to be used by nature to promote survival in contexts other than yeast. Studying worms was also attractive because the worm-aging field had already become a thriving cottage industry.

Moreover, we had recently learned the complete sequence of the *C. elegans* genome, which was the first total genome sequence known of an animal. Although the revolution in genome analyses is best known for decoding the sequence of the human genome, it was the *C. elegans* sequence that proved most critical in this next phase of our research. This trove of knowledge told us that worms had four SIR2-like genes among their collection of 18,000 or so, and provided hope that one or more of them might regulate the life span of these tiny serpentine creatures.

Heidi and I decided to attempt an experiment in worms like the one Matt and Mitch had done in yeast; to determine whether increasing the number of copies of the worm's own

SIR2 genes would extend the life span. But we would do the experiment in a way that was not biased for SIR2; we would ask whether there was *any* worm gene that would increase the life span of these animals when the number of gene copies was increased. The genetic tool of central importance for our experiment was a device called the free duplication. This is a random, small segment of a chromosome that special strains of worms carry above and beyond their diploid set of chromosomes. A strain bearing such an extra piece of DNA will have three copies of all genes present on the duplication instead of the normal two copies; the levels of encoded proteins will thus be increased by 50%. We obtained from the *C. elegans* stock center about 40 of these strains, each with a different duplication. All told, these duplications spanned about half of the *C. elegans* genome, enabling us to sample thousands of genes including the four SIR2 relatives for the potential to extend the life span of worms. Now the experiment of our dreams was simply to discover whether any of these strains live longer than a control strain with no duplication.

The strains of worms with the duplications were placed on petri plates containing worm food (bacterial cells), and their survival was followed over a few weeks by observing whether they continued their characteristic sinuous wriggling across the plates. Animals that stopped moving were prodded with a small spatula to be sure that they had really succumbed. These assays give survival curves shaped like those we have seen for yeast and humans. When Heidi tested the different duplication strains in this manner, there was only one clear winner that stood out by having a life span much longer than normal animals. Remarkably, the duplication in this strain was a piece of chromosome IV that contained the worm gene most closely related to the yeast SIR2, sir-2.1, as well as about 100 other genes from chromosome IV. Even better, other overlapping duplications from chromosome IV that did not contain this SIR2 gene did not extend the life span. This is the sort of step-by-step detective work that has a postdoc scurrying to the lab bench, even on weekends.

To really prove that the extension in life span was due to the sir-2.1 gene itself, Heidi generated animals with extra copies of this gene alone. To do this, she isolated from the worm genome a piece of DNA bearing only the sir-2.1 gene and injected it into the gonads of normal animals. The injected DNA becomes incorporated into the genomes of germ cells. Since *C. elegans* are hermaphrodites, the same animal contains both sperm and egg cells (oocytes) and reproduces by self-fertilization. Thus, progeny resulting from oocytes or sperm that have picked up the sir-2.1 gene will contain this extra DNA in every cell. These offspring are referred to as transgenic animals.

I pestered Heidi every day to find out whether the worms with the extra copies of sir-2.1 were outliving the normal animals. She always made a practice of not telling me how a life span experiment was going until the data set was complete; i.e., every last one of the animals had died. I found this rather frustrating. However, she had a way of tipping her hand: If the result looked encouraging, she would answer that the data were not ready yet, and do so with a smile, but if it did not look promising, she would express irritation at my asking. In this case she smiled, just what I was hoping. Indeed, the worms with extra copies of sir-2.1 lived much longer. This finding brought considerable joy and suggested to me what it must feel like to hit the lottery jackpot. SIR2 genes now appeared to promote longevity in species spanning the billion years of evolutionary distance that separated yeast and roundworms, and it would do so even though the mechanisms causing aging differed. The potential that these genes could be universal determinants of longevity seemed considerable.

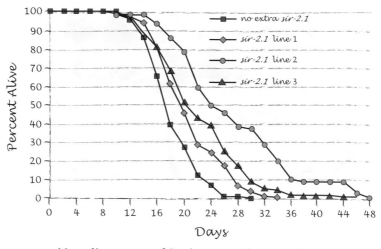

Mortality curves of C. elegans *with extra SIR2.1.*

What does sir-2.1 *do* in worms to promote longevity? Elegant experiments from the labs of Cynthia Kenyon at UCSF and Gary Ruvkun at Harvard over the past eight years have shown beyond any doubt that aging in worms is regulated by a pathway triggered by a worm version of the hormone, insulin. In mammals, insulin circulates in the bloodstream and provides a coordinated signal to many different kinds of cells. Insulin delivers its message by binding to a receptor on the surface of cells. The receptor receives the signal like a radio antenna picking up radio waves and tells cells to carry out specific tasks; for example, taking up glucose from the blood. A defect in insulin signaling leads to an impaired ability of cells to take up glucose, resulting in the disease, diabetes.

In worms, insulin evidently plays another role. Mutations which inactivate the insulin receptor result in a longer life span in worms. Why is this so? Insulin binding to its receptor sets off a cascade of intracellular reactions that comprise a so-called signal-transduction pathway. The net effect of this signaling is to inhibit a protein termed daf-16 that lies at the end of the pathway. Active daf-16 (that is, in the absence of insulin) turns on the expression of a suite of genes that make the worms live longer. Mutating the receptor, therefore, results in a longer life span because it activates daf-16. If the daf-16 gene is first knocked out, then mutating the insulin receptor does not increase the life span. Nature has evidently chosen insulin to tell worms to enjoy a rapid lifestyle of growth, reproduction, and senescence.

Heidi found that the ability of the extra copies of sir-2.1 to extend the life span of worms also required daf-16; in strains in which daf-16 activity had been abolished by mutations, the extra sir-2.1 genes had no effect. We therefore concluded that the way sir-2.1 regulates life span in worms is simple; its normal function is to repress insulin signaling. When more sir-2.1 protein is produced in worms with extra copies of this gene, there is less insulin signaling and the worms enjoy greater longevity. Evolution evidently has coupled the activity of the *C. elegans* sir2-1 gene to the major pathway regulating aging in that organism, the insulin pathway. Meanwhile, I couldn't wait to race to my brand-new department chairman to tell him that we made worms live extra long by giving them more copies of their own SIR2 gene. His response—"just what the world needs, long-lived worms."

Heidi's race to finish these experiments was not strictly due to the usual forces that impel progress in competitive science. She was about to have a baby and wanted to submit the paper before her delivery date. She succeeded in doing this, but just barely, and the child was already out of infancy and on the way to toddlerhood when her research was published in early 2001. The scientific community greeted her paper with enthusiasm in some quarters, but with skepticism in others. The idea that SIR2 genes could extend the life spans of different organisms in which the direct causes of aging were very different was at first not easy to swallow. I told Heidi not to worry about doubting colleagues, because nature often trumps the intuition of the best trained minds.

CHAPTER 8

....

Nature Promotes Survival in the Face of Scarcity

OUR EXPERIMENTS SHOWED the importance of SIR2 genes in yeast and worms. These genes were passed on for about a billion years with the ability to boost the life span well beyond the average. But that's only part of the story. An even closer study of the biology of the SIR2 genes in these two organisms uncovers a fascinating pattern. In yeast, SIR2 genes not only make the mother cells likely to live longer, but also play fundamental roles in helping cells to form spores, the hardiest cell type of all. Spores are specialized cells that can be formed by many microbial species, including yeast, to survive very long periods of time without food and in incredibly hostile environments. They are to cells what Superman is to Clark Kent. In fact, there has been an astounding claim—based on apparently credible research—of the discovery of a bacterial spore that had survived 25 million years!

Spores are constructed from diploid yeast cells, as a part of the sexual cycle, when food is in short supply. Their construction requires an orchestrated pattern of gene expression to build the toughened cellular structure. Success requires that these genes be turned on in a precise order, just as a tasty omelet will only result when the cooking oil is added before the eggs and cheese. Both SIR2 and its very close relative in the yeast genome, HST1, play important roles in the ability of yeast cells to form spores. HST1 (*Homolog of Sir Two*) is also an NAD-dependent histone deacetylase, but it is targeted to regions of the yeast genome different from places where SIR2 is targeted. There HST1 helps coordinate the elaborate order of gene expression necessary to form spores.

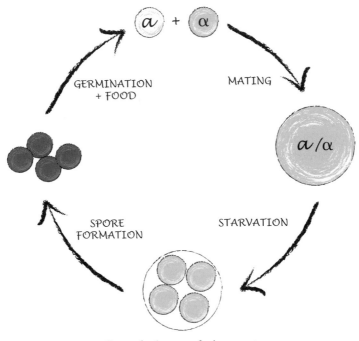

Sporulation cycle in yeast.

SIR2, as we know, preserves the fertility of **a** and α cells by silencing the reserve mating-type genes. This fecundity allows cells to mate and form the fused **a**/α diploid cells that are capable of forming spores. As one might guess, the SIR2, SIR3, and SIR4 genes were all originally identified (in the 1970s by Amar Klar, now at NCI-Frederick; Ira Herskowitz, now at UCSF; and Jim Haber at Brandeis) because mutants in any one of the genes render cells unable to mate and therefore unable to form spores. Thus, SIR2 genes not only trigger longevity in mother cells when times are relatively good, i.e. at least some food is available to allow growth, but also assist the development of the long-surviving spores when times are very lean and growth is not possible.

In *C. elegans* a strikingly similar situation exists. The sir-2.1-regulated pathway of insulin signaling, which determines life span in adults, also controls the formation of a spore-like form in worms. This specialized creature is called "dauer" from the German word for enduring. Dauer animals are formed during larval development and have been studied by Don Riddle at Washington University, Jim Thomas at the University of Washington, and Gary Ruvkun at Harvard, among others. If conditions are good, worms mature to adulthood, but if conditions are poor, i.e., food is scarce, the worms scoot off the normal developmental pathway and form dauers. In other words, the worms enter an alternative developmental track. The dauer animals themselves are tiny, even by roundworm standards, and are hermetically sealed at both ends. They possess large stores of fat, which they burn for energy. Dauers can survive many months while foraging for food, and if they do find enough to eat, they will mature to become normal adults that produce as many progeny as adults that never existed as dauers.

Development of this specialized dauer larval form requires an orchestrated pattern of gene expression, just as yeast spore formation does. It turns out that the insulin pathway regulates this pattern of gene expression and therefore dauer formation, and it does so by activating daf-16. The name daf-16 actually stands for *da*uer *f*ormation, and this gene was identified among mutants that were unable to form dauers. Repression of the insulin pathway when conditions are poor tells worms still undergoing larval development to

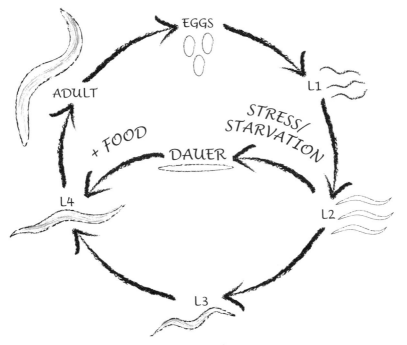

Dauer in C. elegans.

turn on genes for dauer formation. In worms that are already adults, dauer formation is no longer possible, but repression of the insulin pathway still activates daf-16 and makes worms live longer, as discussed in Chapter 6. Thus, by repressing insulin signaling, sir-2.1 not only promotes long life in adults, but also nudges the worms toward the dauer lifestyle when times are lean during development.

Dauer formation in these roundworms is a part of one of nature's most cunning strategies, called diapause, which is found in many insect species. Diapause occurs in response to poor conditions, i.e., cold temperatures or scarce food, and can result in arresting development in larvae, and, in some species, stopping reproduction in adults. In addition to diapause, there are other forms of dormancy in nature. We all know that the bears of Yellowstone Park hunker down in caves when tourists' lunch baskets thin and shadows lengthen. But hibernation also occurs in many other organisms, such as reptiles, amphibians, fish, birds, rodents, and marsupials. There have even been reports of humans surviving extended periods in extreme conditions; for example, children who have fallen through ice into freezing water and are rescued alive up to an hour later. Studies suggest that hibernation puts the brakes on the aging process in rats and hamsters, as measured by cumulative damage to proteins. In the case of Turkish hamsters, longer times of hibernation were also linked to increased life spans. It is interesting that hibernating animals, like dauers, also use fat stores for energy. Garnering energy from burning fat rather than carbohydrates is evidently more compatible with surviving long periods of dormancy, while suspending or slowing down the aging processes.

It is truly impressive that evolution has chosen SIR2 genes for such survival mechanisms, at least in yeast and worms, when they are faced with stress or starvation. SIR2 genes possess an attractive package of features that suit this purpose. They require NAD for their activity, which may make them unusually sensitive to the metabolic activity in cells. Most importantly, they carry out the fundamental enzymatic reaction of histone deacetylation, which regulates the expression state of genes. It remains to be seen whether SIR2 genes play similar roles in higher organisms: promoting longevity in the face of mild scarcity, and providing more specialized mechanisms for long-term survival when the going gets really tough. Evolution's choice to place SIR2 genes in charge of survival in such times of stress squares strikingly well with evolutionary theories of aging, as discussed in the next chapter.

CHAPTER 9

Theories of Aging

SIR2 Is More Evolutionary
Than Revolutionary

HUMANS ARE IN THE PECULIAR PREDICAMENT of observing their slow decline and mortality. This cruel awareness has undoubtedly spurred aspirations to immortality in a next life, which we see etched in the beliefs of major religions. I wondered how our findings about SIR2 and survival might fit in with the long history of ideas about aging. As a molecular biologist, I was concerned that the scientists who have been thinking about aging for many years, mostly evolutionary biologists, would not fully appreciate what our approach could bring to the table. At conferences that include both evolutionary and molecular biologists, people in these two groups tend to hang with their own kind and to distrust the ideas of the other camp. Occasionally the two sides are brought together by circumstance, as in a newspaper article covering one of our studies on SIR2. A well-known evolutionary biologist was consulted for a sound bite and mustered as much enthusiasm as he could stand. He described our work as "relatively harmless." Can the new molecular findings transcend the relatively harmless and add significantly to the more classical ideas and theories about aging?

One of the oldest and most basic ideas about aging is that organisms simply wear out, like 1959 Chevys. This "wear-and-tear" theory undoubtedly does have relevance to certain aspects of aging. The deterioration of blood vessels, hearing, or skin exposed to the sun and wind very likely *is* due to physical or radiation stresses. A popular photograph of a cloistered Tibetan monk in his 80s reveals a smooth face that completely belies his years. But no scientist would seriously believe that this fellow has avoided aging by his sequestered, sedentary lifestyle spent in meditation. The wear-and-tear theory is best viewed as a laudable initial attempt to come to grips with the problem but is not a serious scientific theory; the problem is that people turn out to be more complex than Chevys.

Among the first coherent ideas put forward to explain aging was the so-called rate-of-living theory. This theory focuses on the metabolic rate of animals; i.e., their rate of energy production through respiration by the conversion of oxygen to water. The theory intended to explain why most larger animals, which in general have a slower metabolic rate, live longer than smaller animals, which have a higher metabolic rate. The credo of this theory could be "live fast—die young." Do differences in metabolic rate really cause differences in life spans? The German biologist August Weismann thought so, proposing in the late 19th century that the key factor that "influences the duration of life is purely physiological: it is the rate at which an animal lives." Proponents of this theory concluded that different species of animals are endowed with a relatively constant level of total metabolic output over their lifetime. This theory fits with the more recent observations that metabolism can generate toxic oxygen radicals that damage cells. Studies of certain non-vertebrate animals bred at different temperatures give more ammunition to the proponents. For example, fruit flies have a much longer life span at colder temperatures, which slow down their metabolic rate. Like the hypothesis that telomere-shortening causes aging, the rate-of-living theory is appealing in its simplicity. But is it correct?

Closer scrutiny of the animal kingdom shows that the correlation between life span and total, lifetime metabolism is by no means universally true. Among mammals in particular, bats have comparable metabolic rates to mice, yet live about 10 times longer. This is true even of bat species that do not undergo the apparent restful periods of torpor that we associate with animals hanging upside down from the attic rafters. This means bats burn through 10 times as much oxygen in their lifetime as comparably sized mice. In addition, birds have very high metabolic rates and yet display extremely long life spans for their size.

There is a more fundamental objection to the rate-of-living theory, however. There is no reason that any damage generated by metabolism could not be repaired in cells. Indeed, germ cells are replicated from generation to generation without any aging whatsoever. Thus, organisms ought to be able to evolve life spans that respond to environmental imperatives in a way at least partly independent of their total lifetime metabolism. The omission of this vital consideration limits the usefulness of the rate-of-living theory, although the theory is still instructive in pointing out that metabolism is a relevant piece of the larger puzzle that any incisive model of aging must put together.

J.B.S. Haldane, the Scottish geneticist, had an important insight that would have implications for future aging theories. He pondered why the genetically inherited Huntington's disease, the disease that affected Woody Guthrie, was so frequent. Huntington's disease strikes middle-age adults and leads to the gradual decline of muscular control and death. The mutation for this disease is inherited in a dominant fashion; i.e., the defective gene that causes the disease comes from one parent, whose offspring have a 50% chance of inheriting it. Since this disease is so destructive, why has this gene not been removed from the population by natural selection? The answer is that the symptoms do not begin until later in life, by which time reproduction has occurred and the Huntington's disease gene has already been passed on to the next generation. Haldane's insight was in understanding the fact that natural selection would not be effective at eliminating a gene whose destructive path did not begin until midlife, after reproduction has typically taken place.

In the 1950s, the British immunologist and Nobel laureate, Peter Medawar, applied this simple principle to aging. Medawar reasoned that Haldane's logic applied not only to the gene for Huntington's disease, but to all genes. Therefore, if the function of any gene were to decline in midlife, i.e., after reproduction has occurred, it could not be effectively culled from the population by natural selection. The figure below shows this decline in the ability of natural selection to exert its effects during and after the period of reproduction, leading in turn to a rise in physical or mental defects. By this view, humans are saddled with a genome of 35,000 genes that have not been highly selected for the ideal maintenance of their functions past midlife. It follows that many, many biological processes would fail simultaneously during aging as a passive consequence of the declining efficiency of the genome. This evolutionary theory obviously fits with the widespread deterioration of bodily functions that is part and parcel of normal aging.

An interesting corollary of the Medawar theory was added by George Williams, an evolutionary biologist at the State University of New York in Stony Brook, and bears the ungainly name of antagonistic pleiotropy. He considered this question: What if a mutation occurred that made a gene function better early in life but caused a large decline in its activity later? Of course, the evolutionary theory predicts that it would be selected for and enriched in the population. Is there evidence for genes that exhibit this kind of antagonistic pleiotropy? Williams imagined the process of bone calcification as an example. Any mutation improving bone calcification might be favored as a benefit early in life, for example by

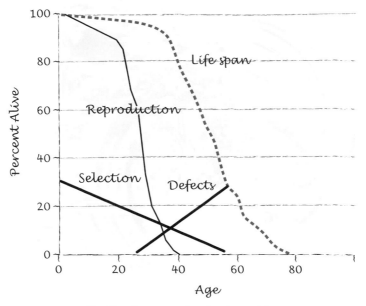

Decline in natural selection with age.

speeding healing of broken bones, even though it could be bad news later by promoting excessive calcium deposition and arthritis. It remains to be seen how frequently antagonistic pleiotropy actually applies to the catalog of human genes, if at all. I consider in a later chapter whether the activity of mammalian SIR2 genes in regulating cell death fulfills Williams's prediction.

But what about the point of view that is completely opposed to those above; that aging is not passive but is genetically programmed? By this view, senescence is advantageous to allow the next generation to flourish without any competition for resources. This kind of "good for the group" idea of aging falls into a category termed group selection. Many evolutionary biologists dislike this idea because a more convincing case can be made that selection acts at the level of the individual rather than at the level of groups. Furthermore, they ask, what if a mutation occurred that abolished an aging program? It would be expected to free up resources that could then be used for reproduction. Such a mutation ought to be enriched by natural selection, meaning that any genetic program of aging would have a very brief shelf life during evolution. As an empirical scientist, I could be swayed into the programmed aging camp if there were convincing evidence for genes that promote aging across species lines. I know of no such genes.

Medawar's evolutionary theory of aging has been bolstered by the notion that there is a trade-off in utilization of resources between reproduction and the maintenance of the soma (the tissues that make up the body excluding the oocytes and sperm). British gerontologist Tom Kirkwood crystallized this idea in a theory of aging termed the disposable soma theory. The idea is that the soma can be discarded after reproduction, because the genes have already successfully been passed on to the next generation. Now imagine a zero sum game between reproduction and somatic maintenance; i.e., increasing resources for one will come at the expense of the other. What factors determine the optimal allotment of resources to these two functions? Mice in the wild have life spans of a few months due to death by predation and exposure to cold. It would make sense for that species to use all of its energies to reproduce as rapidly as possible in this time frame. It is therefore not surprising that mice are fertile one month after birth and have short life spans of two years in captivity where they are protected from predators and cold. Now imagine a bat, which has the added benefit of escaping predators by taking to the air. Since bats are not under the same time pressure as mice, it makes sense that their reproductive schedule is more leisurely and their life spans accordingly longer. The same logic applies to sea turtles, whose hard shells are a serious deterrent to any would-be predator, and which accordingly are very long-lived. Thus, the essential idea of the disposal soma theory is that environmental imperatives shape the tug of war over the available resources between reproduction on the one hand and the maintenance of the soma on the other.

In other words, bats live long because any mutation allowing a more relaxed time frame of reproduction and longer life span would have had an opportunity of being selected for over eons of evolution. In mice the same mutations would have been selected against because they slowed reproduction in the critical first few months of life. An interesting study by Steven Austad at the University of Idaho compared opossums on the Sapelo Island off Georgia with those living on the mainland. Geologists say that this island separated from the mainland about 4,000 years ago, generating the two isolated populations of opossums. Austad found that the island animals, free of many of the predators of the mainland, displayed a longer life span than their relatives on the mainland and also showed slower aging, as measured by the retention of suppleness of their tendons.

Like the rate-of-living theory, this evolutionary theory of aging can explain the rough correlation between body size, metabolism, and life span in animals. Larger animals can better ward off predators, so it makes sense that evolution has selected for a more relaxed reproduction schedule, lower metabolism, and longer life span. Unlike the rate-of-living theory, the evolutionary theory also explains why bats and birds live longer than expected for their body size.

The evolutionary theory implies a very close association between time of reproduction and life span. Just how closely associated are they? Michael Rose at the University of California at Irvine carried out an experiment on fruit flies to test whether slowing the schedule of reproduction automatically triggers a longer life span. Flies have a fertile period in the middle of their life span followed by a time of decline leading to senescence. Rose propagated flies for many generations by withholding mating until the end of their fertile period, i.e., later on average than flies would normally mate. He obtained a stock of flies, which, when left to their own devices, now chose to mate later than the starting strain. Numerous genetic changes are responsible for this alteration in mating behavior. Do these late-mating flies live longer? The answer is a resounding yes. Thus, the relationship between reproductive timing and life span must be a very intimate one. This conclusion is reinforced by another study by Tom Perls at Harvard showing that women who have their first child after age 40 have significantly longer life spans than women in the general population.

Such findings beg the question of how the soma knows that reproduction has occurred. Could there be a kind of a program in which the germ cells (sperm and oocytes) and soma communicate? There are interesting new data suggesting that the germ cells actually send a message to the soma that they are alive and well and, presumably, generating progeny. The message is most likely one or more hormones, although much more work needs to be done to really prove this claim. The best data available come from studies by Cynthia Kenyon at UCSF and Linda Partridge at University College, London, which show that proliferation of oocytes and sperm shortens the life span. The germ cells can be eliminated in developing animals by zapping their gonads with a laser. The laser beam is so thin that it knocks out individual cells like a guided missile. If the germ line is eliminated by zapping the germ precursor cells in developing animals, the life span of the animals when they reach adulthood is lengthened. Moving way up the phylogenetic ladder, there are even studies from certain prisons claiming that the gruesome practice of castration provoked extreme longevity in the affected individuals.

As we can see, the evolutionary theory of aging explains many observations in the natural world and suggests a trade-off between reproduction and maintenance of the soma. But this theory also implies that aging is multifactorial and is influenced by the failure of many genes. The logical conclusion is that there can be no single gene or even a small number of genes that regulates the process. How then can we explain the emergence of SIR2 as a central and conserved regulator of aging by virtue of its ability to promote survival? Well, it's really quite easy, if we suppose that these classical theories have omitted one condition—a harsh environment. For example, a scarcity of food would select for specific mechanisms to slow aging and reproduction until better times arrive.

David Harrison at the Jackson labs in Bar Harbor, Maine, and Robin Holliday, the British biologist who made his name studying recombination, separately proposed this idea in 1989. Harrison described the "enormous selective advantage of mice that could respond to a drought, or other long term condition that severely reduced food supplies, by delaying reproductive senescence." As shown above, delaying reproduction would automatically delay aging. This idea is entirely consistent with the evolutionary theory of aging and is actually a nice capstone to the theory. The evolutionary theory predicts that there is no single gene or even small number of genes that could be altered to make a mouse live as long as a bat. The theory foreshadowed by Harrison and Holliday predicts that a single gene could allow a mouse or a bat to survive much longer in the face of scarcity by delaying reproduction and slowing aging. At this point in time, SIR2 is a good candidate for such a gene.

One more subtle addition to the evolutionary theory seems apt. If nature has designed a regulated process to slow aging in times of scarcity, then the causes of aging must not be infinitely complex and varied. Rather, these causes would all have to respond in lock step when nature pushes the slow-motion button. So this mechanism to slow aging during stress must have evolved to stay one step ahead of aging—whatever the causes of aging happen to be in that organism. Again, SIR2 seems like a good choice for this task because its simple enzymatic activity is very adaptable. In yeast, this deacetylase is targeted to the rDNA to down-regulate that locus. In worms, we imagine that it is targeted to one or more genes in the insulin pathway to down-regulate them.

This rather philosophical chapter brings us back to a fundamental question about aging. Is aging a passive consequence of the lack of selection, or is it a part of a regulated program? My answer would be both. Under conditions of sufficiency, aging plays out passively according to the evolutionary theory with the simultaneous decay of many biological processes. The timing of this process is driven by environmental factors, like the presence of predators, and includes an open line of communication between germ cells and the soma.

However, under conditions of scarcity, a regulated program of survival is kicked into gear that recognizes this environmental imperative and slows aging and delays reproduction. As an NAD-dependent deacetylase, SIR2 may be uniquely equipped to judge whether conditions are flush or lean and to set the pace of aging accordingly. Since SIR2 could extend the life spans of organisms that age for different reasons, it must exert variable functions in different species. As suggested above, nature may have varied the genomic sites where SIR2 is targeted in different organisms. This variation could result in different genes falling under SIR2's sway, leading to different processes being altered. An important question is which environmental conditions of scarcity are able to activate this survival program. In the next chapter, we consider one such condition, calorie restriction.

CHAPTER 10

Calorie Restriction and Life Span

T HE DOGMA THAT WATCHING caloric intake is good for your health has permeated popular culture. Part of this awareness comes from the realization that just like an overabundance of dietary lipids, excess calories will generate unwanted body fat. But we also know that a spartan diet of about 1000 calories per day may make you live longer, even if you are lean to begin with.

Where did this notion come from? Studies done as early as the 1930s demonstrated that lowering food intake could extend the life span of laboratory rodents up to twofold. Careful analysis revealed that the only component in the diet that had to be limited was the caloric content; i.e, the sum of the carbohydrate, protein, and lipids that can be metabolized by cells to generate energy. Calorie restriction (or CR) is effective at any time during the lifetime of the animal. The longer the animals are on the restricted diet, the greater the extension in life span. The diets that are effective in bestowing longevity on these animals are abundant in vitamins, minerals, and essential amino acids but restricted in the total number of calories they provide. This is in stark contrast to the diets of undernourished people in the third world or of individuals with eating disorders. A proper calorie-restricted diet, i.e., one certified by a nutritionist, is almost diametrically opposed to malnutrition.

Lest you consider starting on such a regimen, be warned that a severely calorie-restricted diet has some side effects. The restricted rodents are cold, hungry, small, and lack any sex drive. Also, it is still an open question whether CR would actually extend the life span of humans. It has been noted that the answer to this question does not really matter. Even if a calorie-restricted diet did not actually extend your life span, it would certainly seem like it did! I know of only a few, possibly crazed, individuals who have practiced this kind of diet, including one of the leading scientists who has studied and written about it. Roy Walford subjected himself to this diet as a member of the team of Biosphere II, a private venture in the 1990s, undertaken to see if eight people could sustain themselves in a sealed environment. When the supply of homegrown food did not meet expectations, food restriction became somewhat of a necessity. Another devotee was a student in my summer course on the molecular biology of aging, Douglas Crawford. What the practitioners seem to have in common is a skinny, gaunt appearance and, most importantly, a permanent place of residence in California. The warm and sunny climate may be among the reasons for this geographical clustering.

There is also the problem of self-deception concerning the benefits supposedly conferred by such diets. An 80-year-old fellow from England faxed me a photograph of himself wearing a skimpy swimsuit. He said that he was on a very special diet that made him look half his age. Unfortunately, the fax was barely legible, but contained the promise of a "hard copy to follow" in the mail. I was very eager to read about his diet and told the lab members about this extraordinary individual. One week later the real photo arrived in the mail. He looked exactly like an 80-year-old man in a skimpy bathing suit—so much for the special diet.

A common objection to the significance of the rodent studies is that the restricted animals live longer than the controls largely because the latter group is overfed to the point of ill health. In most of these studies, control animals are allowed to eat as much as they can, a method termed ad-libitum feeding. However, in careful experiments comparing animals that were all on calorically measured diets, Rick Weindruch at the University of Wisconsin found the same trend shown in the earlier studies. Fewer calories gave longer life, even though none of the animals were overfed. CR has a widespread application in triggering longevity, working equally well in mice, rats, and even fruit flies. More recent studies on nonhuman primates are still in progress, and it is too early to know whether CR will provide longevity, but the signs are good. The physiological hallmarks of CR that have been observed in rodents, e.g., lowered body temperature, blood glucose, and insulin levels, are also found in these primates.

The flexible nature of the benefit provided by CR is also impressive. Lab strains of mice typically die of cancers, like thymic lymphoma. A strain of rats called Fischer rats, on the other hand, most often die of kidney disease. Since both mice and rats respond to CR, this means that CR is able to postpone or cure altogether different kinds of diseases in the two organisms. So we see two organisms that die from different causes but whose lives can be lengthened by the same treatment. This is a second example—earlier we saw how SIR2 extends the life span of yeast and worms. It remains to be seen whether SIR2 is also responsible for these effects seen in mammals.

In *C. elegans,* Siegfried Hekimi of McGill University has described a kind of food restriction in worms triggered by mutations that slow their ability to pump in food through the pharynx. These animals, termed "eat" mutants, live longer than controls that pump in food at the normal rate. A second category of worm mutations studied by Hekimi also slow metabolism, but in a different way. They damage the machinery of respiration itself, thereby slowing the rate of oxygen consumption and energy production. These mutations also promote greater longevity.

Scientists have interpreted CR and the effects of Hekimi's slow mutants in the context of the oxidative damage theory of aging. By restricting calories, the fuel for respiration, the generation of oxidative damage should also be reduced. The slow Hekimi mutations will cause the same outcome. It is not surprising that restricted animals or the mutant worms live longer. It is hard to argue with this logic. An old metaphorical device in philosophy called Occam's razor dictates that the simplest explanation of any phenomenon should be presumed to be the correct one. In this example, it would violate the rule of Occam's razor to propose any additional layer of complexity as a part of the essential mechanism by which CR extends life span. However, when experimental observations can no longer be fully explained by the simple model, one must be careful not to cut off one's nose with Occam's razor. Our studies of CR in yeast made us temporarily place Occam's razor back in its sheath.

I asked two of my postdocs, Su-ju Lin and Pierre Defossez, whether we could test if CR worked in yeast. Both Su-ju and Pierre evidently developed good dietary habits in their native countries, Taiwan and France, judging by their svelte, attractive appearances. The time Su-ju spent in my lab was both productive and reproductive. She was very well organized, and gifted as an experimentalist. Su-ju and Pierre completed a series of experiments on CR in yeast, and then she left on maternity leave while we organized the paper. She returned to the lab, the paper was published in *Science* magazine, and she began another very nice set of experiments before again taking a maternity leave. Su-ju returned again to finish those experiments and put together another fine paper for the journal *Nature*.

How could Su-ju and Pierre restrict calories in the diet of yeast? Petri plates that are used to grow yeast typically contain a rich source of nutrients plus the sugar glucose to provide energy. We reasoned that by reducing the glucose concentration from the usual 2%, we would impose a kind of CR. Sure enough, at 0.5% glucose the yeast cells grew a little bit more slowly, and, most importantly, they lived 25–30% longer. Now we could test quickly why the cells lived longer. Was it simply due to a slowing of their cellular metabolism? Or were things more complicated? To wit, we knew that SIR2 promotes long-term survival in yeast and worms in response to a severe scarcity of food. Would SIR2 also be required for yeast cells to harvest the benefit of CR on their life span?

The key experiment was to determine whether a yeast mutant lacking the SIR2 gene would benefit from CR. The experiment gave a clear-cut answer. CR did not benefit the SIR2 mutant. This meant that the extension in life span conferred by CR was mediated by SIR2, presumably because CR triggered an increase in its silencing activity. This hypothesis was strengthened by a further experiment. When Su-ju measured the silencing activity of SIR2 in cells growing in 0.5% glucose, it was indeed more robust. In a parallel set of experiments, Su-ju addressed whether NAD itself was required for the greater longevity in 0.5% glucose. NAD is synthesized by a group of dedicated enzymes in yeast cells. By mutating the genes encoding these enzymes, she was able to lower the amount of NAD in cells without eliminating it, which would have been lethal. The cells with lowered NAD were also unable to enjoy greater longevity on the lean diet.

These experiments told us that CR, at least in yeast, was more complicated than the oxidative damage theory would have predicted. The additional complication was that the extended longevity was not a passive consequence of a lower metabolism but a regulated response that nature has created to sense and respond to scarcity. We might have expected this because of the role of SIR2 genes in survival decisions in yeast and worms. We imagined that CR alters the metabolic

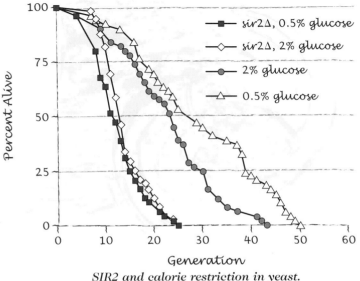

SIR2 and calorie restriction in yeast.

strategy in cells to free up higher levels of NAD that can be utilized by SIR2. This extra NAD increases the silencing activity of SIR2, because it is an NAD-dependent deacetylase, resulting in greater longevity.

It will be important to determine whether the same mechanism applies to animals. *C. elegans* is particularly suitable for these experiments because one can get an answer within weeks instead of years, as would be the case with mice. I managed to convince a new postdoc in the lab, Mohan Viswanathan, to begin these experiments. Mo is not lacking in self-confidence, and he came to the lab with no intention of working on worms. His background was in studying *E. coli* bacterial cells, and he wanted to take the small step up to yeast in my lab. I had to persuade him that studying something as big as a roundworm could be worthwhile. I worked on Mo slowly and gradually, remembering the story of the frog in the bathtub. If the temperature of the water is increased gradually, the frog stays in place, but if it is increased suddenly, the frog leaps out. The key in this case was to provide enough heat to get the juices flowing without scorching the subject. Mo eventually became warmed to study aging in worms instead of yeast.

In the wild, roundworms eat whatever mixture of bacterial cells they happen upon while crawling through soil. In the laboratory, *C. elegans* are typically fed cells of the intestinal bacterium, *Escherichia coli.* The worms snake their way through fields of bacterial cells on petri plates, eating as they go and leaving sinuous trails in their wakes. One problem is that bacterial cells also grow on the petri plates, making it impossible to limit this food supply. Circumventing this problem, however, was relatively easy. We simply mutated the strain of *E. coli* used so that it grows slowly, or not at all. Now by diluting the *E. coli* cells before adding them to the plate, we can give worms a deliberately limited amount of food. Another problem is that worms that are starved too severely will forage, that is, crawl off the petri plates seeking nourishment. It is difficult to do a study on animals that roam the building freely.

We had a clever idea to get around this problem by using mutant worms that are less mobile. These worms, called "roll mutants," cannot crawl at all, but instead roll around on the plates. However, these mutant worms thoroughly foiled us. When they were starved, they rolled off the plates! The instincts of nature were not to be denied. The trick we came upon was to use normal worms, the crawling kind, but not to starve them too severely. This resulted in a substantial extension in their life span.

Other labs have developed ways to restrict worms in liquid culture; i.e., by adding the worms and their food to test tubes with liquid growth medium. After 10 years of yeast life spans, I became such a lover of petri plates that we did not

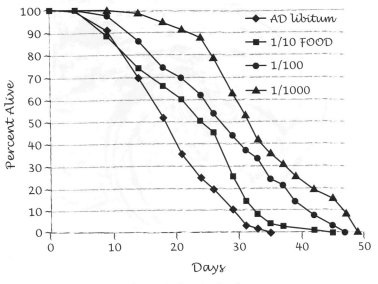

Food restriction in C. elegans.

even attempt experiments using liquid culture. The plate method has the added benefit of being the procedure used in almost all experiments on worm aging I know of in the scientific literature. Why change the worms' media in midstream? But *E. coli* cells are not as defined a food source as the specific ingredients we feed to yeast. How do we know that diluting the bacterial cells does not exert some other important change, in addition to reducing calories? For example, perhaps *E. coli* cells, while sustaining the worms, also contain some substance that is a little toxic to them. In this case, diluting the bacteria may promote longevity in worms simply because it spares them of this poison. These issues will have to be sorted out by researchers over the next few years. Clearly, there is a great deal yet to be learned about SIR2 genes, CR, and life span in worms.

CHAPTER 11

Aging and Cancer in Mammals

I AM OFTEN ASKED WHETHER our results in yeast and worms hold any relevance for humans. One of the main reasons scientists have worked on these two laboratory organisms is that many findings first obtained in these systems turn out to be true of mammals as well. In the case of SIR2, there is a strong evolutionary case to be made that the gene may regulate survival in humans. The logic goes like this: Yeast and worms are very divergent evolutionarily, having separated from some common ancestor about 1 billion years ago, yet both use SIR2 to regulate survival. Any gene whose function has been conserved over that expanse of evolutionary time is very likely to have retained that role during the evolution of mammals, including humans. Evolution does not usually throw away something that serves a useful purpose.

By now we know that SIR2 is a survival factor in yeast and worms, even though the more immediate causes of aging differ in those two organisms. Let's imagine that the insulin-signaling pathway that regulates the life span of *C. elegans* had never been identified previously. The study of sir-2.1 and its regulation of survival in worms would have led to the discovery of this pathway as the function of worm sir-2.1 was probed more deeply. By this reckoning, SIR2 genes are divining rods—they point to the more immediate regulators and, ultimately causes, of aging, because these are the processes that must be brought into play in those tough times of scarcity to promote survival. There is a very interesting implication to this reasoning. A study of the functions of the mammalian SIR2 genes will lead us to immediate causes of mammalian aging, i.e., those processes that limit the life span, such as rDNA circles in yeast and, possibly, oxidative damage in worms. We would then know what we were up against in trying to combat human aging.

In the human genome, there are seven genes with DNA sequences that are similar to the yeast SIR2. By comparing these sequences, computer programs can decide which are most closely related to SIR2 and organize them into a phylogenetic tree. As in a family tree, individuals that are in the same branch are most closely related to each other. More precisely, the physical distance between two genes in the tree, as traced out by moving along the branches, represents how far their sequences have diverged during evolution. Human SIR2-like genes are termed SIRT1–7 where SIRT stands for SIR T(wo). SIRT1 is the most closely related to the bona fide yeast SIR2. Mice have a SIRT1 gene almost identical in sequence to the human. The six other human genes are less closely related and range to the most distant, SIRT6 and 7. Because of the close relationship between SIRT1 and SIR2 itself, it is not surprising that Shin Imai, the postdoc who discovered that SIR2 was an NAD-dependent histone deacetylase, also observed the same activity for SIRT1. But how do mice or humans use this activity? Is it to silence rDNA, as in yeast?

To address these questions, Shin determined where the mouse SIRT1 protein resides in mammalian cells, by using antibodies to pinpoint its location, just as we had in yeast earlier. Recall that in yeast, SIR2 resides at the rDNA, the telomeres, and the extra mating genes. These are the most highly repeated DNA sequences in the yeast genome. We expected

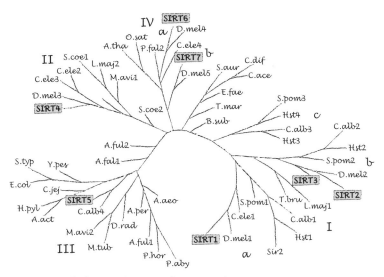

Phylogenetic tree of mammalian SIR2 genes.

SIRT1 to be found at the repeated DNA sequences of mammalian genomes as well; i.e., the rDNA, the telomeres, and the centromeres (the chromosomal region that promotes faithful chromosome partitioning at cell division). Surprisingly, the antibodies showed that SIRT1 resided at none of these locations. It was in the nucleus of cells, but spread out to cover most of the nuclear volume (the space occupied by the cellular home of the chromosomes). Parallel studies showed the same pattern for the human SIRT1. We therefore came to the conclusion that SIRT1 did not concern itself with silencing in the repeated DNA in the mammalian genome. Rather, it was somehow directed to the non-repeated DNA, most likely to regulate the activity of specific genes.

What kinds of genes might SIRT1 regulate? What genes might actually play a role in mammalian aging? Amazingly, recent studies have brought us to the notion that SIRT1 regulates one of the most critical orchestrators of gene expression in mammalian cells, p53. The p53 protein is a tumor suppressor; its presence holds the line against the dark forces of cancer. When cells sustain damage to their DNA, for example by exposure to hydrogen peroxide, which generates oxidative damage, or UV light, which causes radiation damage, p53 springs to action and binds to DNA sites near damage-response genes to activate their expression. This suite of genes then sets off a program to arrest further growth. This arrest can be transient or permanent, in the latter case leading to withdrawal into the dormant state of cellular senescence, much like cultured cells whose telomeres have grown too short.

A second, perhaps more important, way cells call a halt to growth is to commit suicide, a response called apoptosis or programmed cell death. This program is also activated by p53. The important role of p53 in cancer prevention is inescapably underscored by the fact that mutations that cripple the p53 gene are associated with more than 50% of human tumors. Similarly, mice whose p53 gene has been deleted from the genome get cancer at a very high frequency. How the link between SIRT1 and p53 was drawn is a fascinating story.

The background leading to this connection begins with a brief excursion into mouse genetics. Mouse genes can be knocked out in the laboratory by genetic engineering. Mice with a particular gene deleted are called knockout or KO mice. A particular kind of KO mice called "dwarfs" enjoys greater longevity, i.e., a three-year average life span instead of the two years that mice normally enjoy. That extra year would correspond to 30 years in humans. The genetic alteration in these dwarfs affects the pituitary gland and lowers the amount of circulating growth hormone. This, in turn, reduces the amount of insulin-like growth factor 1. We still do not know why the dwarfs are long-lived. It may be because they also exhibit a reduction in the levels of the insulin-like growth factor. Alternatively, since there is a good correlation between body size and life span in nonflying mammals, the longevity of the dwarfs may be an indirect consequence of their small size.

More relevant to SIRT1 and p53 is a study carried out by Pier Guiseppe Pelicci in Milan. His lab generated mutant mice in which a gene called p66shc was knocked out. Amazingly, these mice were both normal in size and extra long-lived, making this the first time that these properties were seen to exist together. Cells cultured from the p66 KO mice differed from control cells in their response to DNA damage. Most importantly, cultured cells of these mice were unable to commit suicide; when treated with damaging agents they just keep on growing. This finding provided the first link between apoptosis and mammalian aging. One imagines that during aging, the progressive loss of cells by apoptosis may lead to organ damage or failure, especially in tissues made up of nondividing cells, like the brain and the heart. By this reasoning, a reduction in apoptosis in the p66 KO mice preserves these organs longer and leads to greater longevity for the animals.

Why do animals have mechanisms of cellular suicide at all? It is now quite clear that growth arrest in the face of DNA damage is an important safeguard against cancer in organs with rapidly dividing cells, i.e., blood, intestine, skin, etc. If damaged cells of these organs were not culled from the animal by apoptosis, they would go on to accumulate mutations each time the cells divided and then progress to tumors. The cell's critical mediator of apoptosis in response to damage is, of course, p53, which can sense the damage and tell cells it is time to commit suicide before it is too late. Harking back to George Williams, the roles of p53-dependent apoptosis in preventing cancer and, possibly, causing aging may be an excellent example of antagonistic pleiotropy—a process that is beneficial early in life but harmful later on. Cancer surveillance is obviously critical in helping animals to progress into adulthood and to reach their time of reproduction. Natural selection would thus smile on it. Continued apoptosis later in life, however, may well bring down entire organs and contribute to aging. As we have seen before, things that happen late in life escape the forces of selection.

I wrote an article that went with the Pelicci paper in *Nature* in 1999, in a section of the magazine titled "News and Views." These short articles accompany larger research papers and attempt to place them into a broader context for the readers of the journal, including nonbiologists. I discussed how p53-dependent apoptosis may limit life span and that p53 KO mice might also live longer than controls, except for the fact that they get cancer early in life. This raised in some minds the specter of a rather unpleasant trade-off between aging and cancer; i.e., preventing the former might foster the latter or vice versa. A recent paper from the lab of Larry Donhower at Baylor actually gave testimony to this trade-off. His group generated a genetically altered strain of mice with a hyperactive p53 gene. These mice are virtually cancer-proof due to the enhanced surveillance, but their organs deteriorate at an early age, leading to a shorter life span. In truth, I now find the idea of some sort of devil's bargain between aging and cancer as counterintuitive and unlikely to be correct. Since the prevalence of cancer is known to rise dramatically with age, Occam's razor would dictate that any genetic alteration that slowed aging would also slow cancer; i.e., the two phenomena would pull together rather than at cross purposes.

One way to invalidate the idea of a trade-off would be to construct a situation to interfere with the bad apoptosis, the kind that might be related to triggering aging, without damping the good kind, which protects against cancer. In fact, the p66 KO long-lived mutants demonstrate this property; it seems clear that they *do* retain intact cancer surveillance mediated by p53, since they do not get tumors early in life. Evidently removing p66 from the picture alters p53 in a subtle way, reducing the unwanted apoptosis associated with aging but leaving cancer surveillance intact. Therefore, there is no obligatory trade-off between aging and cancer. The fact that calorie restriction promotes longevity and also reduces cancers further reinforces this conclusion.

Now we finally come to the relationship between p53 and SIRT1. It turns out that p53 becomes activated in a very interesting way when cells sustain DNA damage. The protein is tagged with acetyl groups upon its activation, the same acetyl tags that mark the histones near actively expressed genes. This exciting discovery by Wei Gu, at Columbia University, was the first example of a regulatory protein other than histones that was acetylated. After we published our findings that SIR2 proteins were NAD-dependent histone deacetylases, he called our lab with an idea. Perhaps, in addition to deacetylating histones, SIRT1 also deacetylated p53, converting the active, acetylated p53 to a less active, deacetylated form. Such an activity would mean that SIRT1 was a repressor of p53; i.e., it damped the activity of the tumor suppressor. This would endow SIRT1 with a very provocative function in mammals; it would reduce p53-dependent apoptosis in response to DNA damage. Just as in yeast and worms, SIRT1 would promote survival, in this case of challenged mammalian cells.

We thus entered a productive collaboration with Wei Gu, the success of which demonstrates the power of modern communication networks. As of this writing, I still have never met Wei Gu. All of the many exchanges we have had in successfully working together were by telephone, fax, and E-mail. We quickly found that the purified SIRT1 protein could deacetylate p53 in a test tube, and very efficiently. We also found that SIRT1 could deacetylate p53 in cultured mammalian cells. The key question was whether SIRT1 could dictate the decision to live or to die in mammalian cells. Two kinds of experiments told us that yes, indeed it could. First, giving cells more of the SIRT1 protein rendered them more resistant to apoptosis when challenged with damaging agents. We did this experiment by inserting the SIRT1 gene into a virus that injects its DNA stably into the genome of mammalian cells. When our genetically altered virus infected the cells, it brought in the SIRT1 gene, which integrated along with the viral DNA into the genome. Resulting cell lines had extra copies of the SIRT1 gene and expressed higher levels of the protein. The fact that these cells were more resistant to apoptosis perfectly fit the notion that SIRT1 negatively regulated p53.

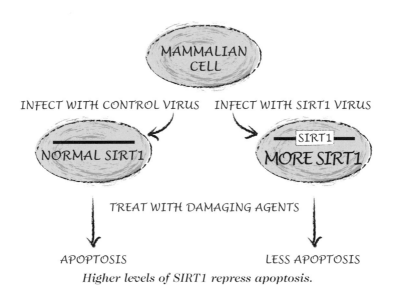

Higher levels of SIRT1 repress apoptosis.

In a second experiment, we did exactly the opposite. Instead of giving cells more SIRT1, we interfered with the activity of the cells' own SIRT1. It seemed to make sense that decreasing the activity of the cells' own SIRT1 would give rise to a hyperactive p53 and that these cells would be more prone to apoptosis when challenged. This was also found to be the case. In this experiment, we inhibited the cells' own SIRT1 in two different ways. One way was using a chemical inhibitor, nicotinamide, which is the half of NAD that is attached to the ADP-ribose. By binding to SIRT1, nicotinamide prevents the deacetylation reaction from occurring. Without the ability to deacetylate, SIRT1 can't do its job. The other method employed a mutant version of SIRT1 itself as an inhibitor. This mutant cannot carry out the deacetylase reaction, but will get in the way of the cells' own functional SIRT1. The mutant SIRT1 gene was introduced into cells by the virus and expressed at high levels for this purpose. Homayoun Vaziri of the Whitehead Institute, who also participated in these studies, developed this latter trick.

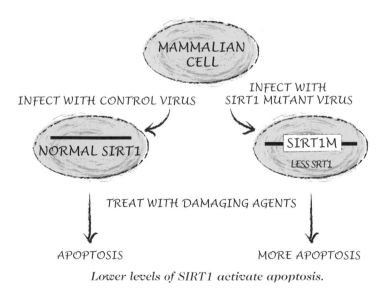

Lower levels of SIRT1 activate apoptosis.

There now seems to be little doubt that SIRT1 down-regulates genes under the control of p53 and that this activity strategically influences what the cells' response to DNA damage will be. The most obvious way SIRT1 could do this is by deacetylating p53. There is a second possible mechanism that may also be important. By binding to p53, SIRT1 could piggyback to sites near the suite of genes controlled by the tumor suppressor. Once near these genes, SIRT1 could down-regulate them by deacetylating the nearby histones. By this second model, the relevant activity of SIRT1 is histone deacetylation, and the deacetylation of p53 itself could be but a sideshow, an indirect consequence of p53's intimate association with the SIRT1 deacetylase.

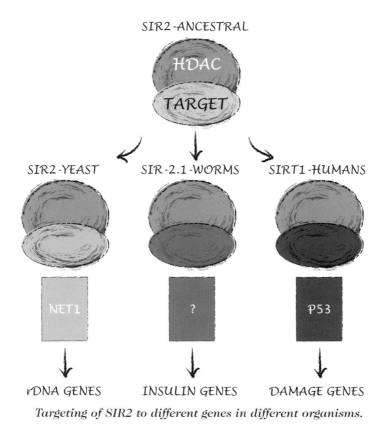

Targeting of SIR2 to different genes in different organisms.

I personally like this latter model because it explains neatly how SIR2 could regulate different processes to slow aging in different organisms. In yeast, SIR2 is targeted to the rDNA by a DNA-binding protein called NET1, where it exerts its effects on life span. In worms, we figure that some other DNA-binding protein targets sir-2.1 to genes of the insulin-signaling pathway, where it regulates dauer formation and life span. In mammals, yet another DNA-binding protein, p53, targets it to genes that regulate apoptosis. SIR2 has evidently evolved to recognize different DNA-binding partners in different organisms. In this way, the SIR2 deacetylase can adapt to regulate different suites of genes and functions and to remedy whatever processes are the immediate causes of aging.

As of now, there is no hard evidence as to whether increasing the activity of SIRT1 causes cancer. Two observations, however, make me suspect that it would not. First, no human SIR2 gene has been identified among the many genes that can cause cancer—the so-called oncogenes. Since oncogenic mutations work by increasing or altering the activity of a host gene to cause cancer, we might have expected SIR2 to be guilty of such mischief. Of course, the skeptic can say that we still have not analyzed all human tumors and SIR2 genes may yet appear as oncogenes. Second, the behavior of the p66 KO mice suggests that there is a window of "smart" inhibition of p53, which will not increase the potential for cancer even as it promotes longevity.

It is always a temptation to think that the latest findings will be the final word on any scientific problem. In the case of mammalian SIRT genes, there are reasons to be very wary indeed. We do not yet know how many functions SIRT1 has in mammalian cells. The regulation of p53 and apoptosis just happens to be the first one identified. For all we know, there may be many other SIRT1 partners besides p53, and the regulation of apoptosis may be but one of many SIRT1 functions. Moreover, there exist six other SIR2-like genes in the mammalian genome. One or more of them may carry out functions that have an effect on survival. We simply cannot yet assess the importance of apoptosis per se in what might be a sea of biological activities of SIRT genes.

Finally, a little humility is also in order. We have traveled down a very narrow path of SIR2 genes in our studies of aging. Different roads followed by other labs may lead more directly to the fountain of youth. The role of circulating hormonal factors in coordinating survival in animals is another contender among processes deserving further study. In particular, the roles of mammalian insulin and the insulin-like growth factor must be carefully examined. In addition, the area of cross talk between the germ cells, i.e., oocytes and sperm, and the soma is extremely fertile for study. The hormones that might mediate such cross talk could well be a major focus of the drug companies in their search for an anti-aging "potion." We are clearly at the point when we need hard answers to the question that started this chapter: Do the genes and mechanisms that have been linked to aging in lower organisms also influence the process in mammals, and, if so, how? These answers will likely unfold over the next five years or so.

CHAPTER 12

Relevance to Human Aging

ONE OF THE QUESTIONS I ASK myself is—What if SIRT genes turn out *not* to be relevant to human aging? Sure, there have been other molecular models of aging that have fallen by the wayside. In fact, most of them have. I can always go back to the grocery store where I developed stock-boy skills in high school and college. However, getting up at 7 a.m. every day to stock shelves no longer holds the same appeal. But let's hold the prophets of doom at arm's length for the time being, and contemplate strategies that can test whether SIRT genes really do determine mammalian life span. Such cogitation suggests three ways to test whether SIRT genes really matter. These are genetic studies in mice, human genetics, and drug discovery. Each has its pros and cons.

Genetic Studies in Mice

We have already seen that mouse genes can be knocked out in the laboratory. This was made possible by the culturing of mouse embryonic stem (ES) cells from early embryos. These cells are pluripotent; i.e., they can transform themselves into any kind of mouse cell, skin, heart, brain, etc., given the right cues. The gene of interest can be knocked out using recombinant DNA technology in these cultured ES cells. When the KO cells are injected into a normally developing mouse embryo at an early stage in development called blastocyst, they can differentiate along with their neighbors into many body parts. If they happen to differentiate into sperm or egg cells, the resulting adult animal will transmit the knocked-out gene when mated, thereby deleting the gene in the progeny.

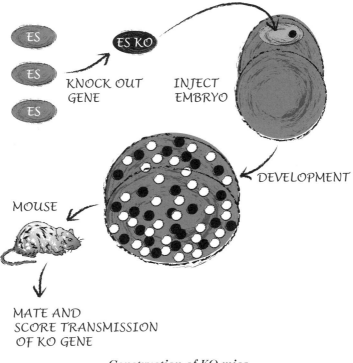

ES

ES

ES

ES KO

KNOCK OUT
GENE

INJECT
EMBRYO

DEVELOPMENT

MOUSE

MATE AND
SCORE TRANSMISSION
OF KO GENE

Construction of KO mice.

It turns out that extra gene copies can also be added to make transgenic animals, just as in yeast and worms. These genetic manipulations allow us to test the hypothesis that mammalian SIRT genes regulate survival. Let us take as an example the mouse SIRT1 gene. The hypothesis that SIRT1 has a crucial impact on survival would lead us to expect that knocking out the gene shortens the life span and adding extra copies increases it. The beauty of this kind of study is that it can provide evidence that SIRT1 determines life span in a mammal. Not everyone, however, would agree with the significance of this on its own. The skeptic could still say, OK, SIRT genes regulate life span in mice, but what about humans? Furthermore, studying mice is excruciatingly slow compared to yeast or worms. Generating and breeding KO mice takes one to two years. Measuring how long they live takes another two to three years.

Another concern is that there are seven SIRT genes to worry about. It is common in the mammalian genome for genes to be present in multiple copies. In these cases, each family member is usually expressed in only a subset of the different tissue types—one gene may be expressed predominantly in the liver, but another in the brain. Our preliminary studies on the mouse SIRT genes do confirm differences in tissue distribution. If specific SIRT genes are doing the job in different organs, it may be necessary to increase expression of all of them to cover the vital organs and make the whole animal live longer.

Human Genetics

In the past decade or so, it has been possible to identify many genes responsible for certain inherited diseases. The task of the disease-gene hunters is a laborious one. They study families in which some members have inherited the disease and others have not. This level of analysis usually can determine the inheritance pattern. For example, cystic fibrosis is an autosomal, recessive trait. This means that the gene resides on one of the 22 chromosomes, excluding the X and Y, and individuals with the disease, in a very unlucky twist of fate, have inherited a defective copy from both parents. We have entered a new era of human genetics as a result of a recent technological revolution. The entire genomes of diseased and healthy individuals can now be compared using genetic sequence features, or markers, that vary in the population as a whole. You might have the DNA sequence AAGCTTAA at a given position on the chromosome, and I might have AAGGTTAA at the same location.

If this DNA sequence lies very close to a gene, it and the gene will travel together and be linked from generation to generation. This is because the process of recombination, which would split apart the DNA sequences to either side of the chromosomal break, is unlikely to occur in such a small region. These DNA sequence markers often differ between the matched pair of chromosomes that we have inherited from Mom and Dad. The gene hunters search for a very close association between the inheritance of a disease and a given set of linked markers. Such an association pinpoints the location of the disease gene to the chromosomal region defined by the markers. This exercise frequently narrows a disease gene to a region containing anywhere from about 10 to 100 genes. Zeroing in from there on the correct one involves DNA sequencing to locate gene mutations—or studies of the biology of the genes in the region—or both. A little luck doesn't hurt either.

In the case of SIRT genes, we start off one jump ahead of the disease-gene hunters. The good news is that we already know what genes we are interested in, SIRT 1–7, now called candidate genes. The bad news is that these genes may have nothing to do with human life span (boy, was that tough to say!). Having candidate genes greatly simplifies the task of checking for any association with a disease, or, as in this case, with aging. One must first identify genetic differences in SIRT genes in the general population. Best are single-nucleotide polymorphisms, or SNPs. These are differences of 1 nucleotide in the DNA sequence, for example AAGCTTA versus AAGGTTA. In general, SNPs occur at about 1 per 700 nucleotides, so the SIRT genes, which contain about 2000–3000 nucleotides of coding DNA, would have, on average, 2–4 SNPs. Let us say there were two different SNPs for SIRT1 in the general population, symbolized as **A** and **a**. You or I would have possible genotypes on our chromosomes of **AA**, **Aa**, or **aa**. Tests to rapidly determine these genotypes from blood samples could be routinely developed, and you and I could be classified.

Now comes the hard part. We are not looking for a disease with obvious and unmistakable symptoms, like cystic fibrosis, but have the more daunting task of studying variations in aging and life span. What would we do next? A good option would be to access tissue samples from large-scale studies that have focused on the health of many individuals over several decades. These include, for example, the Baltimore Longitudinal Study and the Framingham Heart Study. We could then look for an association of a SIRT1 genotype with long life. Let's say the frequency of the **aa** genotype in the general population is 1 in 100, but in people who live to be 90 or over it is 1 in 10, or higher. We might conclude the **aa** genotype is associated with long life span, and SIRT1 may well determine longevity in humans.

There are limitations to this kind of study. There almost certainly exist many genetic factors that determine whether someone lives to be 90. These include avoidance of genes that would predispose toward the common causes of death: cardiovascular disease, cancer, stroke, etc. Thus, the ability of the **aa** genotype to cause long life is at the mercy of many other variables, and any effect of this genotype on longevity could easily be masked. In other words, people are not like inbred strains of mice, worms, or yeast. Rather than relying on life span, a better approach might be to look at other, more narrow properties of the individuals. For example, perhaps we would find that high retention of cognitive ability would correlate with **aa** and suggest a role of our gene in the aging

AA *Aa* AA AA
aa Aa aa AA
Aa aa
aa Aa Aa Aa   GENERAL POPULATION
Aa Aa *Aa*
Aa Aa Aa
AA *ad* Aa AA Aa
AA *Aa aa* Aa
aa

Aa AA *aa*
aa aa aa
aa aa aa Aa   LONG-LIVED PEOPLE
Aa *aa aa*
aa aa aa Aa
Aa *Aa* Aa AA
ad Aa AA *aa*
AA *aa* Aa

Association of a SIR2 genotype with longevity.

of the brain. In a perfect world, we would test how fast individuals in the study were aging biologically compared to their chronological age. With such biomarkers of aging, this could even be done with mid-lifers instead of the elderly. Sadly, there is still no such assay that is any more reliable than that first impression we get in casting our eyes on someone.

In addition, we would have to be careful that the long-lived individuals were not unique in some other aspect that indirectly affects their SIRT1 genotype. For example, if all of the 90-year-olds happened to have family origins in Sardinia, they may have an unusual SIRT1 genotype because of their isolated geographic origin; i.e., because they have not been a part of the common gene pool. Their longevity may be due to some other factor unrelated to SIRT1. An even bigger problem is that there is a pretty good chance that any given gene, like SIRT1, will not have any useful SNPs at all. That is, we may all have the same sequence of the DNA that encodes the SIRT1 protein. In this case, there will be no opportunity to compare individuals with differing genotypes, and this approach falls flat.

So, the big pros in the human genetics approach are that it is potentially rapid and it is the real deal, humans instead of mice. The potential cons are many, but the SNP approach certainly ought to be tried.

Drug Discovery

We live in a world that scarcely could have been imagined even a century ago, with drugs apparently available for every conceivable malady. Most drugs are smallish organic molecules (molecular weight less than 2000 daltons) and found in nature or synthesized chemically. How are these new drugs discovered? Typically, a pharmaceutical company decides on a "target." Targets are usually proteins of sufficient biological interest that companies will invest time and money in identifying drugs that bind to them. To be approved as prescription drugs, candidate compounds must receive approval of the Food and Drug Administration (FDA) for safety and efficacy. Beyond prescription drugs lies a vast array of products at the local health food stores, some of which claim to slow aging and improve health. These claims are universally unsubstantiated.

Why are there no prescription drugs for aging? Aging has never been considered a disease, so our entire biomedical infrastructure has no provisions for developing and testing authentic palliatives. But common sense suggests that the same deterioration of tissue with age we see on the outside must occur internally as well, leading to organ failure and diseases. Furthermore, aging has not been understood at a detailed enough level to spur attempts to develop medicine cabinet treatments. However, this latter impediment is fading away as our knowledge of the aging process and the regulation of survival deepens. Does this mean we *are* on the verge of a new era of drugs to make us live longer? Probably, but let's first consider how a drug could demonstrate whether SIRT genes regulate mammalian aging.

Let us suppose we decide to develop a drug to alter the activity of SIRT1. One approach would be to use the most up-to-date knowledge about the SIRT1 protein to help in the design of a drug. The three-dimensional structures of two of the SIR2 proteins have been determined to a resolution of 1.5 angstroms, or one-tenth of one one-billionth of a meter. The exact positioning of the NAD molecule can be seen. With this kind of detailed picture, chemists can begin to imagine the exact composition and structure of small molecules that would bind to SIRT1 and alter its activity. To date, unfortunately, this model-based approach of drug discovery is far from routine. However, the strategy will increase in importance as new, powerful, computational methods gain center stage in biology.

Another, more classical approach is to use the known activity of SIRT1, the NAD-dependent deacetylase, as a guide in a large-scale, so called high-throughput, screen for drugs. With the aid of robotics, one can test small organic molecules from diverse collections compiled by the pharmaceutical industry for their ability to increase or decrease SIRT1 activity in the test tube. Hundreds of thousands of compounds can then be screened over a period of months to identify the rare one that influences the activity of SIRT1, the target of the screen.

Any compound obtained in this way would have to meet several additional criteria in order to be developed into a drug. It would have to be relatively stable in the bloodstream

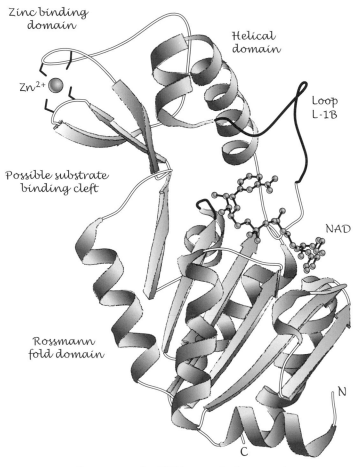

Zinc binding domain

Helical domain

Zn²⁺

Loop L-1B

Possible substrate binding cleft

NAD

Rossmann fold domain

N

C

Structure of a SIR2 protein from archaea.

and efficiently taken up by cells. It would also have to be free of any toxic effect on the patient and effectively treat a disease. So the reality of screening is that only something like 1% of the initial candidate compounds result in drugs that reach the marketplace. SIRT1 poses one additional problem.

PLASTIC WELLS WITH SIR2   LIBRARY OF COMPOUNDS

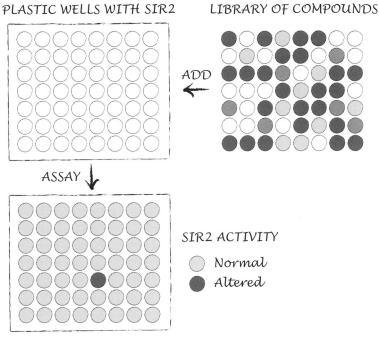

ADD ←

ASSAY ↓

SIR2 ACTIVITY

◯ Normal
⬤ Altered

High through-put screen for drugs affecting SIR2.

Most drugs work by inhibiting the activities of their targets. To extend life span, we need a drug that binds to SIRT1 and *increases* its activity. This may be yet another hurdle, since many believe that it is easier to find inhibitors than activators, although this is more an article of faith than an empirical fact.

If we had a drug in hand, we would test whether SIRT1 regulates mammalian life span. We would feed it to mice and see whether they live longer. Of course, similar tests could be done on dogs, cats, and, eventually, humans. The power of this method depends on the drug's selectivity in hitting its target. If the drug only interacted with SIRT1, of all the tens of thousands of proteins, lipids, carbohydrates, and nucleic acids in a mouse, we could be confident that its biological consequences result from an alteration in SIRT1 and only SIRT1. If the drug could also bind to other cellular constituents, its effects may have nothing to do with SIRT1. In this case, unpredictable consequences to the patients could result. Therefore, the more selective the drug, the better.

I began to wonder about the possibilities of developing drugs for aging about seven years ago. This led me into an interesting cul de sac—starting a biotech company. The story of that journey seems worth telling.

CHAPTER 13

Starting a Biotechnology Company

Drugs for Aging?

I HAD AN EARLY INDOCTRINATION in biotech start-ups. As mentioned earlier, I worked in the lab of Mark Ptashne as a postdoctoral fellow at Harvard in the late 1970s and early 1980s. The Harvard Biological Laboratory was a building of great history and minds, with an entrance flanked by two bronze rhinoceri. They served, one assumed, as a reminder to students and postdocs to beware of getting gored. The place was buzzing with the notion that genetic engineering would usher in a new era in the pharmaceutical industry, and many entrepreneurs were born, at least in their own minds.

The pioneering companies Genentech and Amgen had recently come into existence. Mark worked toward starting his own company during those years, a process that involved meeting business people of many different stripes. One day those high officials who managed Harvard's endowment approached him about the possibility of Harvard backing the company. I recall one meeting of interested parties from both coasts convened in the O'Hare Hilton. The official in charge of investing Harvard's billions told us that this venture in all probability would make everyone in the room millionaires. This was heady stuff for a postdoc living on about 10K a year.

But Harvard soon became concerned that any involvement in a commercial venture would tarnish its image and pulled out of the deal.

Meanwhile, Mark would go out and meet with new potential investors on a regular basis. He would return saying that he was offered a deal an order of magnitude better than the last one. If this were literally true, he would soon corner all of the financial resources of the planet. Soon the company, called The Genetics Institute, was officially launched by Mark and his colleague Tom Maniatis. The company did well and was eventually acquired by American Home Products.

I ended up not joining this venture, but was a founder in a separate start-up involving several postdocs and faculty at Harvard. This was not a case of politics, but an opportunity to be a bigger fish, if in a smaller pond. Our company, BioTechnica International, taught me many lessons about what to do and not to do to develop a successful company. We had a fantastic scientific staff, especially in the area of genetic engineering of agricultural crops, but little business experience. This latter deficit was never made up, eventually spelling doom for the company. The remnants after the wheels came off were absorbed into a midwestern company that sold seeds to farmers; a sorry end to our labor of love.

Around eight years ago, I served as a hired gun to help a firm of financial investors evaluate a possible investment. Venture capitalists, or VCs, search for innovative young companies in which to invest and grow their capital. For this purpose, a VC from a company called Polaris Bioventures, Terry McGuire, and I spent a day at a small start-up company sandwiched in tight quarters in the garment district of New York. Terry was a rising star in the VC world and projected an aura of ambition and drive. On the flight back to Boston, I broached the subject of starting a company in the area of aging. This chat began a regular series of discussions between us that spanned several years. The critical issue was deciding when the time was right to begin trying to raise funds. Investors like to see a clear business plan, a portfolio of patents (this they call intellectual property or IP), and good people. I had no business plan and a total of only one patent on our methods to discover longevity genes in yeast, but I was eager. The vision was to develop drugs that alter the activity of important targets affecting aging.

Terry introduced me to Cindy Bayley, who held a Ph.D. in virology and an MBA, both from the University of Chicago. Cindy was a driving force in the biotech start-ups launched by ARCH Venture Partners, a venture capital fund that began as an offshoot of the University of Chicago technology licensing office. Her stock in trade was to evaluate any new proposals for start-ups that might appear on ARCH's plate and to launch new companies. She and her VC comrade, Bob Nelsen, had put together a long string of successful companies. She had been thinking about a start-up in aging for some time. A few weeks before Christmas in 1997, on the day of the lab holiday party at my house, Cindy came to meet me. She arrived in my office wearing a sport coat, her hair flying out in many directions. Her sentences came out fast and animated and she spoke with her arms flailing. She was equally comfortable discussing science, business, or, for that matter, opera, and I could tell that she was evaluating me, a process the business folks call "due diligence." I took her to the party to meet my lab members and my son. She clearly liked the technology and decided to work on forming an aging company full time.

Not long thereafter, I also met Bob Nelsen. His appearance belied his record of accomplishments as a VC; Bob appeared to be about 18 years old, prompting me to wonder if he had already found the fountain of youth. His crisp, clear thinking would be a guiding strength behind the company. He and Cindy said that I needed to increase the base of the venture by finding a scientific partner, preferably Gary Ruvkun or Cynthia Kenyon, two leaders in studying aging in *C. elegans*. Gary, an old Harvard pal, was somewhat hesitant, because at that time he was more interested in investigating diabetes than aging. Cynthia was someone I had also known since I was a postdoc at Harvard, when she was a graduate student, and we had already talked previously about her ideas for starting a company. And Cynthia was eager. She was also very much in the public eye as an accomplished researcher on aging, having appeared in outlets ranging from public television to *Glamour* magazine. The fact that she also looked half her age did not discourage media attention.

After much back and forth, Bob Nelsen and Terry McGuire made us an offer for a first round of financing. It was surprisingly meager, and we turned it down. We were not enthusiastic about trying to realize our vision of drug discovery on the cheap. In retrospect, I think the VCs were just having a little fun with us. Nonetheless, Cindy, Cynthia, and I had our company incorporated, if not financed. Around the same time, we started making our pitch to other VC firms across the country. The reaction of these firms fell cleanly into two categories. One group loved the concept and thought it was innovative in an area that would only increase in importance with an aging population.

The others just didn't get it. They fixated on the idea that aging is not a disease and therefore not an appropriate subject for a biotech start-up. In addition, they offered the practical objection that there is currently no mechanism for getting FDA approval for an anti-aging drug, because FDA approval requires showing efficacy in the treatment of a disease. Personally, I think that aging will eventually be classified as a disease. Even if this does not happen, an anti-aging drug may well slow down conventional illnesses of aging, like osteoporosis or prostate cancer. An indication of an anti-disease property may well come from testing in special strains of rodents that mimic specific human diseases. This path would then lead to human trials. Thus the first bona fide anti-aging drug may come out of clinical trials for the treatment of a disease like osteoporosis.

What amazed me about these dog and pony shows to secure start-up money was that each presentation was before only one or two VCs who were charged with making final decisions for their firms. One of these visits was at Oxford Bioventures in Boston, where we presented to their lead VC, Jonathan Fleming. He towered above the rest of us at 6' 5", and projected largeness in other ways as well, most importantly as a financial expert who got in at the ground floor. Oxford had a long-standing interest in aging—they got it. Years earlier, their eminent VC Alan Walton assembled the financial backing that launched Geron. Jonathan said that he loved us, but I would also have to make our pitch to Alan, whom he referred to as "the mountain." I talked to Alan, and the mountain was evidently still searching for the fountain. Oxford and Arch were both very interested. Meanwhile, Polaris lost interest in us, and in biotech altogether, because they were then faring spectacularly well by investing in Internet companies. A Seattle group that sometimes co-invests with Arch called Perennial Partners was also brought into the syndicate. Dave Maki from Perennial served as a valuable legal counsel and negotiator of deals.

There was another kind of interest out there, as well. I got a call from an investment banker in San Francisco to set up a meeting. He had been approached by a family that wanted to invest 25 million dollars in a company to work on developing drugs for aging. This was much more money than the aggregate sum the VC groups would offer, and for the same fractional ownership of the company. These wanna-bes of high finance outside the VC world are called angel investors. Why didn't we take their manna from heaven? Cindy convinced us that the hidden assets of VC backing are invaluable to the success of start-ups. They have many contacts with other companies and boundless experience to bring to the table. Furthermore, their backing is an imprimatur that will attract more investors as the company grows. Finally, they are ready to invest again in a later round of financing. It occurred to me they may also have a realistic view on how long it really takes to succeed, which may not be true of an aging billionaire who is trying to keep his/her young spouse happy. So we went the VC route, getting a much better deal than their first offer—now 8.5 million for a little more than 50% of the company.

Next it was important to position the major chess pieces for the company, the CEO, CSO, CFO, COO, in other words, all the C something Os. I recall a 4-way phone conversation, another staple in the business world, in which Jonathan raised the name Ed Cannon for CEO. I perked up because I knew Ed from his days on the Brandeis faculty going back 20 years or so. More recently, I knew Ed from my time on the scientific advisory board of DYAX Corp., where he was director of research and development. Ed is a fine, widely read scientist with ample business experience, and the idea of working closely with him was extremely attractive. I told Jonathan I would arrange a meal with Ed, who subsequently told me—over our power lunch—that he would consider the position. He immediately grasped the science and its implications for a successful start-up. Our VCs and Ed reached agreement and he was on board.

This hiring, incidentally, fixed the location of the company to the Boston area. We reside at One Kendall Square, arguably the biotech Mecca of the world and close to MIT. Ed proceeded to hire Pete DiStephano, an experienced group leader at Millennium Phamaceuticals, as our CSO. The filling of other staff scientist and scientific advisory board positions has also begun. It is fascinating to evaluate young candidates for staff positions. Only a small number of all Ph.D. graduates have the entrepreneurial spirit that translates into a burning desire to work in a biotech start-up. When you find people like this, you do not want them to escape before signing on the dotted line.

We needed to negotiate a deal with MIT to license my patents, now grandly totaling three. This moved ahead, with MIT obtaining an ownership stake in the process. Any concerns about conflict of interest, which shadowed the earlier days of university–biotech relationships, had evidently vanished. But Cindy was not satisfied to have only my intellectual property under the tent and began negotiating with several other lab chiefs and their home universities to license in more IP. This is a good business strategy because it creates more research and development possibilities in house. It also reduces the number of possible competitors launching their own companies.

The next vital issue, which continues to this day, was the company name. I had the idea of Elixir Pharmaceuticals, which was not exactly an epiphany. The first meaning of "elixir" in *The Times English Dictionary and Thesaurus* is "an alchemical preparation supposed to be capable of prolonging life." Cynthia did not like the name because it sounded too much like Exelixis, a more mature biotech company in California that, although not working on aging, also works on worms. My feeling was that I had no idea what Exelixis meant, so I was not too concerned about any possible confusion. Nonetheless, we did some canvassing and came up with alternative names. We considered Fountain, but thought it was too bland. We were struck by Ichor, the blood that flowed in the veins of the gods giving them immortality. Then Cindy found in a reference lexicon that the word has an obscure, more mundane meaning that would not behoove us, i.e., pus. So we met with a high-powered design firm who would develop our logo and also help with a name. After weeks of deliberations, they came back with a Letterman-like top-10 list of candidate names, the leader of which was VIAX. We quickly moved on down the list. We are still called Elixir.

As a member of the Board of Directors of Elixir, I sit in on business meetings every few months. The business people arrive at the table and promptly display their cell phones, pagers, and E-mail devices, the last of which are preferably Blackberrys. Lacking as colorful a plume, I display my Bic pen and occasionally, for variety, a pack of lozenges. Usually at least one of the board members "attends" by conference call. The meeting is an orderly discussion by voices near and far about pressing matters, usually involving the financing and business development of the company. This is what was missing at my earlier start-up, BioTechnica. We also talk about the latest science and the research directions of the company. At the end we leave the room feeling good, and, if the stars are aligned properly, proceed to a fine restaurant for dinner.

A major organizational concern for the academic entrepreneur is how to separate university and company business. MIT has no problem with their faculty members starting companies. In fact, they strongly encourage it because it serves two purposes. It provides a path for licensing the patents from MIT labs, which of course MIT owns. The more patents that get developed in companies, the greater the chances for downstream royalties, which are split between the inventors and MIT. Moreover, start-ups give company founders a second source of income. This means MIT does not need to worry about paying us so little as faculty. The head of MIT's technology licensing office, Lita Nelsen, played a role in trying to marry me to possible investors while I was beating the bushes for financing. The fact that both she and Bob Nelsen spelled their names with an E seemed like a good omen to her.

A more sinister conflict may occur between the company and one's own lab, however. How do you separate what work goes on in the lab at MIT versus what goes on in the company? I still do not know the answer to this question, but I am optimistic that the two can be kept separate and complementary rather than competitive. This is because the goals of the research lab and those of the company are different. My lab tries to learn more and more about the basic biology underlying aging and survival. The company, meanwhile, is trying to translate this knowledge into drug development. This process begins with screening up to a million compounds for effects on a target protein, say SIRT1. A compound that alters the activity of a target would then be tested for life span effects in worms and mice. Any "hit" passing these tests would then be suitable for chemical modifications that improve its potential properties as a drug candidate in humans. This tedious, arduous path would not suit most academic scientists, even if they did have the resources to execute the plan, which they don't.

In an ideal world, the new information produced in academic labs will feed into drug development strategies for Elixir. In return, any compounds that the company finds may prove useful additions to the research in the university labs. For example, the drug could finally show that SIR2 proteins determine mammalian life span. In Japan there has traditionally been a much closer association between industry and academia. The two have also cooperated closely in the U.S. in some fields, for example, chemistry. In the biological sciences, this has not traditionally been the case, but the increasing importance of biotechnology in our society is bringing universities and companies closer together than ever before. This trend can only continue as health sciences assume an ever more important role in the 21st century.

It will be interesting to see whether a spate of anti-aging start-up companies now start appearing. Recently, a company called Centegenetix was set up by two Harvard researchers to follow up on their studies of centenarians, i.e., people who live past 100. They have identified families with higher than expected numbers of long-lived individuals, suggesting a genetic basis for their longevity. The flagship enterprise of their company will be attempting to identify the genes promoting this longevity. I have also heard of a few companies just getting off the ground that will focus on calorie restriction. I would guess that they are being sold to investors as the only companies focusing on a regimen that is actually known to work in mammals. Finally, the trailblazer Geron is still alive and kicking but has shifted its emphasis from telomeres to stem cell therapy, as discussed in Chapter 14.

An important question for anti-aging companies is how the large pharmaceutical industrials will respond to them. Ideally, close cooperation between large pharmaceutical companies and anti-aging start-ups would accelerate the progress of research and development. But big pharma is typically very conservative and likely to view anti-aging companies with a skeptical eye, at least initially. I have a little first-hand knowledge of this because of my chaired professorship. My official title is Novartis Professor of Biology, which means that the Swiss pharma giant—Novartis—paid MIT to establish the chair. As a result, representatives from Novartis pay a courtesy call once or twice a year. When I told their liason, Joerg Staeheli, about Elixir, he became very intrigued and invited us to their headquarters in Basel.

The day we spent there was great fun; Joerg took us to their roof deck where we could view three European countries with the naked eye. But it was clear that at least some of their scientists could not get past the apparent hurdle that aging cannot be considered a disease. Nevertheless, lunch in their plush dining room featured a nice red wine produced at Novartis's own vineyards in Spain. Now that's class! A major goal thus remains—convincing big pharma that there is a new area of human health and product development that they will need to be a part of. If biotech start-ups do their jobs properly, especially by generating promising new compounds that could lead to drugs, this transition should happen soon enough.

CHAPTER 14

Aging Gracefully in the 21st Century

T HE MOST COMMON RESPONSE I hear after I tell new acquaintances what I do for a living is "work fast." But there are many of us who do not heed what we already know about our everyday habits and how they affect our time on this earth. Probably the most effective way to limit your lifetime and damage your health in the process is to smoke cigarettes. This does not stop millions from taking up this habit, especially in China and other Asian countries. Whenever I visit Europe, it is apparent, even in that bastion of the developed world, that smoking is much more widespread there than in the States. Another simple way to reduce your quality and quantity of life is by a bad diet. There are more obese people in the U.S. than ever before. Although it may be debated whether calorie restriction will prolong human life span, there is little doubt that a gluttonous diet of junk food will shorten it.

In addition, there is the even more hotly contested subject of exercise. Evidence from laboratory rodents shows that exercise will cause a very small increase in life span, but nowhere near the extension observed by restricting calories. I am asked often why exercise is not a *bad* thing, since it will increase respiration and the production of those pesky oxygen radicals. I answer that this logic is too simplistic; exercise changes our metabolism in more ways than merely an increase in respiration during the workouts. The net effect of exercise on the aging process is therefore complex and not easily predictable. The benefits to health and quality of life, however, are unquestionable.

I exercise three or four times a week at the local gym, either running for aerobics or lifting weights for muscle tone. I am not certain whether it will make me live longer, but I know it makes me feel good. One must again be on guard, however, for the trap of self-deception. I was working out at a machine for my biceps, when a pretty young woman in spandex walked over. I thought she might want some tips about techniques to achieve a youthful, sculpted physique. Instead she said "Excuse me, 'sir,' when will you finish on that machine?" I somehow did not think her choice of appellation was in recognition of my research renown.

We are bombarded with claims that health supplements will fight diseases of aging and lengthen our life span. As I said earlier, there is presently no credible basis for these claims. One category of these supplements is vitamins. In general, moderate doses will do no harm and can actually compensate for inadequate diets that are deficient in essential vitamins. I do worry about people who take mega-doses, both for their health and their pocketbooks, since much of the excess material is not absorbed and quickly passes out of the body. I take a multiple vitamin that contains the Bs, a moderate dose of C, a low dose of E, and folic acid. Again, the feeling is that the vitamins can't hurt and the folic acid may help fight cardiovascular disease. Another interesting supplement with reported benefit for cardiovascular disease is red wine. This one is an easy choice.

There is also a school that strongly believes in hormone supplements. The idea is that we can benefit by supplementing certain hormones that the body stops producing as we grow older. Examples include growth hormone, the steroid hormone DHEA (dehydroepiandrosterone), estrogen, and melatonin. There is some evidence to suggest that growth hormone can increase muscle mass in aging individuals, although whether this is a good or bad thing for longevity is not known. Trials of the hormone DHEA are in progress, but I would be surprised if any benefits are found. Estrogen stops being made in postmenopausal females. Its supplementation is recommended by the medical community and is of clear benefit to some tissues, for example, bone. There may also be hazards, since the hormone has been associated with increased risks for breast cancer. Early claims connecting melatonin and aging are mistaken, although the drug is evidently effective for jet lag. The decrease of all of these hormones in the body is likely an indirect consequence of aging rather than an actual cause of it. Therefore, supplementation should not affect the underlying aging process, although some Band-Aid effects on certain organs are possible.

Will we witness pharmaceutical products to slow aging in our lifetimes? I would say so. I figure that the first products rationally developed to slow aging will appear in the next 10–20 years. Some would question the ethics of interfering in the normal aging process. This viewpoint misses the mark for two reasons. First, we are not talking about making people live to be 500. The increase likely to be delivered by any new drug should, at best, be similar to what calorie restriction can do, that is, a 20–30% extension in life span but without the associated misery. This is still nothing to sneeze at, especially if robust health accompanies the extra years. Thus, there is no fear that the menacing specter of immortality will haunt anti-aging drugs. Moreover, we have already passed through a century in which life span in developed countries has nearly doubled. Although much of this effect is due to reduced infant and childhood mortality, the life expectancy of adults has also increased. Surely, drugs that would combat aging and diseases of aging are in the best spirit of biomedical research.

Another highly promising area for the future is body-replacement parts. If your liver begins to fail, someone will grow a new one in the lab to replace it. The hot button area of biology for this task is stem cell research. Stem cells are precursor cells, i.e., the progenitors that can be induced to grow into many different kinds of differentiated cell types. Embryonic stem (ES) cells that can develop into any tissues are obtained from early embryos. As discussed in Chapter 11, the mouse variety is routinely used to generate KO animals.

In humans, ES cells were first cultured only a few years ago and have generated a storm of excitement and controversy. The excitement is that these cells may eventually be a source of replacement organs. One example under current investigation is whether human ES cells can slow or even halt the progress of Parkinson's disease when injected into the affected area of the brain. The controversy comes from the fact that these cells are obtained from human embryos, providing a possible medical use and market for aborted fetuses. This concern has been partly diffused because there are tens of human ES cell lines already in existence. These cells can be grown indefinitely in petri plates, providing, in principle, an inexhaustible supply. President Bush's position put forward in the summer of 2001 found an apparent middle ground that using existing cell lines was permissible but deriving new lines was not. However, it remains to be seen whether these existing lines are made readily available to researchers, since at least some are in the hands of private companies. Moreover, growing these cells is tricky, because they can easily lose their potential to transform themselves into new cell types. One worries that some or all of the "available" cell lines have already been botched.

There is also another class of stem cells, which is found in adults. Unfortunately, it is doubtful that these adult stem cells are as plastic as ES cells in their potential to differentiate. There is also the problem that stem cells themselves seem to deteriorate with age. Nonetheless, adult stem cells could provide a bonus because they would create a perfect tissue match for everyone. Any kind of organ transplantation can be rejected by the immune system of the host if there is not a close match. What must be matched between donor and recipient are certain proteins on the surface of cells, or else the body's immune system will recognize the transplanted tissue as foreign and destroy it. Adult stem cells get around this problem because the donor and recipient are one and the same person! If you need a new kidney, someone will obtain a sample of your own stem cells and grow you a rejection-proof kidney. Of course, our understanding of how to get cultured stem cells to grow into specific organs is still very primitive. It may take decades before we can generate high-quality organs for transplantation.

There is a second way to get a perfect tissue match involving another controversial frontier of biology called cloning or nuclear transplantation. This method was most famously used to make Dolly the sheep. Ian Wilmut at the Roslin Institute in Scotland took a cell from an adult sheep and placed its nucleus into a sheep oocyte whose own nucleus had been removed. The embryo containing the transplanted nucleus developed into a sheep genetically identical to the original donor. I have been told that Dolly has a personality different from any other sheep, but this is not a consequence of nature but of nurture. She gravitates to people and cameras in a way that makes her herd-mates blush. More recently, cloning by nuclear transplantation has also succeeded in mice, cattle, and even cats.

It is hotly debated whether human cloning should be allowed as another alternative for infertile couples or couples who want to duplicate a lost child. The discussion can be waged on two levels, the practical and the ethical. Practically, it turns out that many of the cloned animals die before they are born or soon after they are born, or are deformed in some way as adults. This technical limitation is a disqualifier for human cloning, even without getting into the ethical issues.

However, for one of the most important applications of cloning, concerns about minting new people need not apply. Let us say we wish to generate ES cells that are a perfect match to the donor. One could take a cell from the donor and inject its nucleus into an oocyte whose nucleus had been removed. One possible source of such egg cells would be fertility clinics, although this source is clearly quite limited. The resulting clone need only divide to the 100-cell stage before tissue-matched ES cells can be harvested, cultured, and grown into an organ. This use of the method has been termed "therapeutic cloning." So, if you need a new liver, someone could clone you, but only to the 100-cell stage as a source of matched ES cells, which could then be grown in culture.

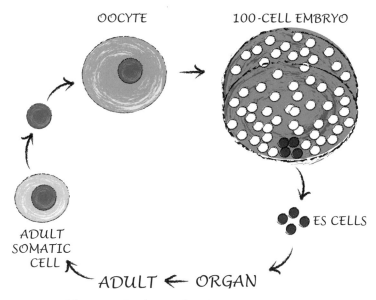

OOCYTE   100-CELL EMBRYO

ES CELLS

ADULT
SOMATIC
CELL

ADULT ← ORGAN

Therapeutic cloning for tissue regeneration.

Some would still have misgivings using ES cells obtained from nuclear transplantation or, for that matter, from aborted fetuses. Others are against cloning, period. I think that the latter group was incited further by the choice of the term "therapeutic cloning" to describe the use of nuclear transplantation to obtain stem cells. This process is not really cloning at all, since the transplanted oocyte is smaller than the head of a pin when its stem cells are obtained. A recent claim that might calm some is that oocytes can be tricked into initiating the early stages of embryonic development using their own nuclei, i.e., without any need for nuclear transplantation. This procedure, called parthenogenesis, would guide embryos to the 100-cell stage, by which time their stem cells can be obtained, and skirt many of the ethical issues surrounding ES cells. It would fail, however, to provide a perfect tissue match for the patient, except for the case in which the egg cell is obtained from the patient herself.

Slowing aging without simultaneously improving the quality of life would be a Pyrrhic victory, an empty triumph. The optimistic school believes that affecting the core aging mechanisms would necessarily mitigate diseases of aging. In such a case, the additional years of life would be accompanied by good health. Given that a wide spectrum of diseases in rodents can indeed be slowed or prevented altogether by an anti-aging regimen of calorie restriction, I tend to agree with this view. But even if some diseases remain stubbornly resistant to interventions that slow aging, other, more directed medical advances are on the horizon. Cures for cancer will be especially important, since the incidence of the disease rises so steeply with age. It makes sense to me that some relatively common cancers, like prostate cancer, are intimately tied to aging itself and would therefore respond beneficially to anti-aging drugs. Therapies for other cancers will continue to be found on a case-by-case basis. We can also expect progress in treating the other major diseases, as basic knowledge about molecular causes of these diseases gets translated into therapies. History books a millennium from now may record this period as the golden era of health advances.

As for me, I plan to glide through the rest of the 21st century the way I finished off the 20th; lots of rock music, a good bottle of wine, the daily newspaper, and a ready book. It wouldn't be bad to settle down again, either, maybe at age 100. I expect more surprises on the research front from my lab, and don't plan to stop studying aging, at least for a while. There is even a modest level of hope of improving my golf swing. It will be fun to be around—as long as possible.

Snapshots

The photographs on the following pages show some of the people who carried out experiments described in this book. They are for the most part snapshots taken in the lab or at lab events at various times over the period covered by the book.

David Lombard   *Nick Austriaco*   *Tod Smeal*

Su-Ju Lin   *Brad Johnson*

Shin Imai and wife, Toshiko

Mitch McVey

Lenny Guarente

Mark Ptashne

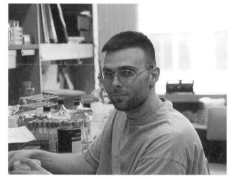

Matt Kaeberlein

Heidi Tissenbaum

Brian Kennedy

David Sinclair   Kevin Mills

Bibliography

Chapter 1

Finch C.E. 1990. *Longevity, senescence, and the genome.* The University of Chicago Press, Chicago, Illinois.

Gompertz B. 1825. On the nature of the function expressive of the law of human mortality, and on a new mode of determining life contingencies. *Philos. Trans. R. Soc.* **115**: 513–585.

Hayflick L. 1994. *How and why we age.* Ballantine Books, New York, New York.

Mortimer R. and Johnson J. 1959. Life span of individual yeast cells. *Nature* **183**: 1751–1752.

Shay J.W. and Wright W.E. 2000. Hayflick, his limit, and cellular ageing. *Nat. Rev. Mol. Cell Biol.* **1**: 72–76.

Chapter 2

Griffiths A., Miller J., Suzuki D., Lewontin R., and Gelbart W. 1993. *An introduction to genetic analysis*, 5th edition. W.H. Freeman, New York, New York.

Hartwell L., Hood L., Goldberg M., Reynolds A., Silver S., and Veres R. 2000. *Genetics: From genes to genomes.* McGraw-Hill, Boston, Massachusetts.

Chapter 3

Harley C.B., Futcher A.B., and Greider C.W. 1990. Telomeres shorten during ageing of human fibroblasts. *Nature* **345**: 458–460.

Herskowitz I., Rine J., and Strathern J. 1992. *The molecular and cellular biology of the yeast* Saccharomyces. *2. Gene expression* (ed. E. Jones et al.), pp. 583–656. Cold Spring Harbor Laboratory Press, Cold Spring Harbor, New York.

Kennedy B.K., Austriaco N.R., Jr., Zhang J., and Guarente L. 1995. Mutation in the silencing gene SIR4 can delay aging in *S. cerevisiae. Cell* **80:** 485–486.

Lewin B. 1997. *Genes VI.* Oxford University Press, Oxford, United Kingdom.

Shore D. 2000. The Sir2 protein family: A novel deacetylase for gene silencing and more. *Proc. Natl. Acad. Sci.* **97:** 14030–14032.

Watson J.D., Hopkins N.H., Roberts J.W., Steitz J.A., and Weiner A.M. 1987. *Molecular biology of the gene*, 4th edition. Benjamin Cummings, Menlo Park, California.

Chapter 4

Guarente L. 1997. Link between aging and the nucleolus. *Genes Dev.* **11:** 2449–2455.

Harlow E. and Lane D. 1999. *Using antibodies: A laboratory manual.* Cold Spring Harbor Laboratory Press, Cold Spring Harbor, New York.

Kennedy B.K., Gotta M., Sinclair D.A., Mills K., McNabb D.S., Murthy M., Pak S.M., Laroche T., Gasser S.M., and Guarente L. 1997. Redistribution of silencing proteins from telomeres to the nucleolus is associated with extension of life span in *S. cerevisiae. Cell* **89:** 381–391.

Chapter 5

Blasco M.A., Lee H.W., Hande M.P., Samper E., Lansdorp P.M., DePinho R.A., and Greider C.W. 1997. Telomere shortening and tumor formation by mouse cells lacking telomerase RNA. *Cell* **91:** 25–34.

Bodnar A.G., Ouellette M., Frolkis M., Holt S.E., Chiu C.P., Morin G.B., Harley C.B., Shay J.W., Lichtsteiner S., and Wright W.E. 1998. Extension of life-span by introduction of telomerase into normal human cells. *Science* **279:** 349–352.

Gottlieb S. and Esposito R.E. 1989. A new role for a yeast transcriptional silencer gene, SIR2, in regulation of recombination in ribosomal DNA. *Cell* **56:** 771–776.

Kaeberlein M., McVey M., and Guarente L. 1999. The SIR2/3/4 complex and SIR2 alone promote longevity in *Saccharomyces cerevisiae* by two different mechanisms. *Genes Dev.* **13:** 2570–2580.

Sinclair D.A. and Guarente L. 1997. Extrachromosomal rDNA circles—A cause of aging in yeast. *Cell* **91:** 1033–1042.

Smeal T., Claus J., Kennedy B., Cole F., and Guarente L. 1996. Loss of transcriptional silencing causes sterility in old mother cells of *S. cerevisiae*. *Cell* **84:** 633–642.

Strehler B. 1986. Genetic instability as the primary cause of human aging. *Exp. Gerontol.* **21:** 283–319.

Szilard L. 1959. On the nature of the aging process. *Proc. Natl. Acad. Sci.* **45:** 30–45.

Chapter 6

Brachmann C.B., Sherman J.M., Devine S.E., Cameron E.E., Pillus L., and Boeke J.D. 1995. The SIR2 gene family, conserved from bacteria to humans, functions in silencing, cell cycle progression, and chromosome stability. *Genes. Dev.* **9:** 2888–2902.

Frye R. 1999. Characterization of five human cDNAs with homology to the yeast SIR2 gene: Sir2-like proteins (sirtuins) metabolize NAD and may have protein ADP-ribosyltransferase activity. *Biochem. Biophys. Res. Commun.* **260:** 273–279.

Guarente L. 2000. Sir2 links chromatin silencing, metabolism, and aging. *Genes Dev.* **14:** 1021–1026.

Imai S., Armstrong C.M., Kaeberlein M., and Guarente L. 2000.

Transcriptional silencing and longevity protein Sir2 is an NAD-dependent histone deacetylase. *Nature* **403**: 795–800.

Strahl B.D. and Allis C.D. 2000. The language of covalent histone modifications. *Nature* **403**: 41–45.

Stryer L. 1995. *Biochemistry.* W.H. Freeman, New York.

Tanny J.C., Dowd G.J., Huang J., Hilz H., and Moazed D. 1999. An enzymatic activity in the yeast Sir2 protein that is essential for gene silencing. *Cell* **99**: 735–745.

Chapter 7

Branicky R., Benard C., and Hekimi S. 2000. clk-l, mitochondria, and physiological rates. *BioEssays* **22**: 48–56.

Friedman D.B. and Johnson T.E. 1988. Three mutants that extend both mean and maximum life span of the nematode, *Caenorhabditis elegans*, define the age-1 gene. *J. Gerontol.* **43**: B102–B109.

Harman D. 1957. Aging: A theory based on free radical and radiation chemistry. *J. Gerontol.* **2**: 298–300.

Kenyon C., Chang J., Gensch E., Rudner A., and Tabtiang R. 1993. A *C. elegans* mutant that lives twice as long as wild type. *Nature* **366**: 461–464.

Kimura K.D., Tissenbaum H.A., Liu Y., and Ruvkun G. 1997. daf-2, an insulin receptor-like gene that regulates longevity and diapause in *Caenorhabditis elegans*. *Science* **277**: 942–946.

Klass M. 1977. Aging in the nematode *Caenorhabditis elegans*: Major biological and environmental factors influencing life span. *Mech. Ageing Dev.* **6**: 413–429.

Melov S., Ravenscroft J., Malik S., Gill M.S., Walker D.W., Clayton P.E., Wallace D.C., Malfroy B., Doctrow S.R., and Lithgow G.J. 2000. Extension of life-span with superoxide dismutase/catalase mimetics. *Science* **289**: 1567–1569.

Riddle D.L., Blumenthal T., Meyer B.J., and Priess J.R., Eds. 1997. *C. elegans II*. Cold Spring Harbor Laboratory Press, Cold Spring Harbor, New York.

Tissenbaum H.A. and Guarente L. 2001. Increased dosage of a *sir-2* gene extends lifespan in *Caenorhabditis elegans. Nature* **410**: 227–230.

Wallace D.C. 1999. Mitochondrial diseases in man and mouse. *Science* **283**: 1482–1488.

Wilson J.D., Foster D., Kronenberg H., and Larsen P., Eds. 1998. *Williams textbook of endocrinology,* 9th edition. W.B. Saunders, Philadelphia, Pennsylvania.

Chapter 8

Chu S., DeRisi J., Eisen M., Mulholland J., Botstein D., Brown P.O., and Herskowitz I. 1998. The transcriptional program of sporulation in budding yeast. *Science* **282**: 699–705.

Lyman C.P., O'Brien R.C., Greene G.C., and Papafrangos E.D. 1981. Hibernation and longevity in the Turkish hamster *Mesocricetus brandti. Science* **212**: 668–670.

Riddle D.L. and Albert P.S. 1997. In *C. elegans II* (ed. D.L. Riddle et al.), pp. 739–768. Cold Spring Harbor Laboratory Press, Cold Spring Harbor, New York.

Tauber M.J., Tauber C.A., and Masaki S. 1986. *Seasonal adaptations of insects*. Oxford University Press, New York.

Chapter 9

Austad S.N. 1997. *Why we age: What science is discovering about the body's journey through life*. John Wiley & Sons, New York, New York.

Guarente L. 2001. SIR2 and aging—The exception that proves the rule. *Trends Genet.* **17**: 391–392.

Harrison, D.E. and Archer J.R. 1989. Natural selection for extended longevity from food restriction. *Growth Dev. Aging* **53**: 3–6.

Holliday R. 1989. Food, reproduction, and longevity: Is the extended lifespan of calorie-restricted animals an evolutionary adaptation? *BioEssays* **10**: 125–127.

Hsin H. and Kenyon C. 1999. Signals from the reproductive system regulate the lifespan of *C. elegans*. *Nature* **399**: 362–366.

Kirkwood T.B. 1977. Evolution of ageing. *Nature* **270**: 301–304.

Kirkwood T.B. 1999. *Time of our lives: The science of human aging*. Oxford University Press, Oxford, United Kingdom.

Medawar P. 1946. Old age and natural death. *Mod. Q.* **1**: 30–56.

Rose M.R. 1991. *Evolutionary biology of aging*. Oxford University Press, New York, New York.

Sgro C.M. and Partridge L. 1999. A delayed wave of death from reproduction in *Drosophila*. *Science* **286**: 2521–2524.

Williams G. 1957. Pleiotropy, natural selection, and the evolution of senescence. *Evolution* **11**: 398–411.

Chapter 10

Lakowski B. and Hekimi S. 1998. The genetics of caloric restriction in *Caenorhabditis elegans*. *Proc. Natl. Acad. Sci.* **95**: 13091–13096.

Lin S.J., Defossez P.A., and Guarente L. 2000. Requirement of NAD and SIR2 for life-span extension by calorie restriction in *Saccharomyces cerevisiae*. *Science* **289**: 2126–2128.

Masoro E.J. 1993. Dietary restriction and aging. *J. Am. Geriatr. Soc.* **41**: 994–999.

Weindruch R. and Walford R.L. 1988. *The retardation of aging and disease by dietary restriction*. C C. Thomas, Springfield, Illinois.

Weindruch R., Walford R.L., Fligiel S., and Guthrie D. 1986. The retardation of aging in mice by dietary restriction: Longevity, cancer, immunity and lifetime energy intake. *J. Nutr.* **116:** 641–654.

Chapter 11

Brown-Borg H.M., Borg K.E., Meliska C.J., and Bartke A. 1996. Dwarf mice and the ageing process. *Nature* **384:** 33.

Frye R.A. 2000. Phylogenetic classification of prokaryotic and eukaryotic Sir2-like proteins. *Biochem. Biophys. Res. Commun.* **273:** 793–798.

Gross A., McDonnell J.M., and Korsmeyer S.J. 1999. BCL-2 family members and the mitochondria in apoptosis. *Genes Dev.* **13:** 1899–1911.

Luo J., Nikolaev A.Y., Imai S., Chen D., Su F., Shiloh A., Guarente L., and Gu W. 2001. Negative control of p53 by Sir2α promotes cell survival under stress. *Cell* **107:** 137–148.

Migliaccio E., Giorgio M., Mele S., Pelicci G., Reboldi P., Pandolfi P.P., Lanfrancone L., and Pelicci P.G. 1999. The p66[shc] adaptor protein controls oxidative stress response and life span in mammals. *Nature* **402:** 309–313.

Vaziri H., Dessain S.K., Ng Eaton E., Imai S.I., Frye R.A., Pandita T.K., Guarente L., and Weinberg R.A. 2001. hSIR2[SIRT1] functions as an NAD-dependent p53 deacetylase. *Cell* **107:** 149–159.

Vogelstein B., Lane D., and Levine A.J. 2000. Surfing the p53 network. *Nature* **408:** 307–310.

Chapter 12

Maxwell R.A. and Eckhardt S.B. 1990. *Drug discovery: A casebook and analysis.* Humana Press, Clifton, New Jersey.

Perls T.T. and Silver M.H. 1999. *Living to 100: Lessons in living to your maximum potential at any age.* Basic Books, New York, New York.

Chapter 13

Werth B. 1994. *The billion-dollar molecule: One company's quest for the perfect drug.* Simon and Schuster, New York, New York.

Chapter 14

Blau H.M., Brazelton T.R., and Weimann J.M. 2001. The evolving concept of a stem cell: Entity or function? *Cell* **105:** 829–841.

Campbell K.H., McWhir J., Ritchie W.A., and Wilmut I. 1996. Sheep cloned by nuclear transfer from a cultured cell line. *Nature* **380:** 64–66.

DePinho R.A. 2000. The age of cancer. *Nature* **408:** 248–254.

Lovell-Badge, R. 2001. The future for stem cell research. *Nature* **414:** 88–91.

Glossary

Acetylated protein: a protein with a small molecule called an acetyl group attached to it.

ADP-ribosylation: attachment of ADP-ribose to a protein.

Aging: deterioration of an organism over time.

Amino acids: subunits from which proteins are made. There are 20 different types of amino acids, and different proteins consist of different sequences of these 20 types.

Antagonistic pleiotropy: any process that is beneficial early in life (before reproduction) and harmful later in life (after reproduction).

Apoptosis: the process of cellular suicide in response to cellular damage or other signals.

ATP (adenosine triphosphate): an important energy compound in metabolism.

Cell line: cells multiplied from a single cell removed from an organism and propagated in culture.

Cellular senescence: elderly condition of cells in culture, when after many rounds of division, they have ceased to divide further.

Centromere: region of DNA that promotes faithful separation of **chromosomes** after they have been copied during cell division.

Chromatin: stretch of DNA wrapped in histones.

Chromosomes: large DNA molecules capable of copying themselves.

Deacetylation: the *removal* of an acetyl group from a protein, usually by an enzyme called a histone deacetylase, for example, the enzyme SIR2.

Differentiation: process by which pluripotent cells turn into specialized cell types such as neurons, muscle, etc.

Diploid: organisms (or cells) containing two copies of each chromosome.

Enzyme: protein that stimulates a chemical reaction, e.g., adding an acetyl group to a protein.

Gene expression: process by which a gene directs the synthesis of its product, usually RNA, which in turn is translated into a protein.

Gene silencing: process by which the DNA in a region of a chromosome is rendered inaccessible or "silent" by **deacetylation** of histones. Both **gene expression** and **recombination** are generally inhibited in these regions.

Genome: the entire complement of genes of a given organism.

Histones: proteins around which DNA is wound in a cell. A stretch of DNA wrapped in histones is called **chromatin**.

Life span: survival time of an individual (i.e., how long he/she lives). More specifically: **average life span** is the mean survival time of individuals in a given species or population; **maximum life span** is the survival time of the longest-living individual recorded within a given species or population.

Longevity: the life span of an individual.

Lysine: a specific amino acid found in proteins to which an acetyl group can be attached.

Mitochondria: compartment within the cell where oxygen is consumed and ATP is produced, in a process called respiration.

NAD (nicotinamide adenine dinucleotide): molecule in cells required for both metabolism and deacetylation of proteins by SIR2.

Nucleolus: region within the cell's nucleus where the ribosomal DNA is located, and where ribosomes are made.

Oxidative damage: damage to DNA, RNA, proteins, etc. caused by oxygen radicals.

Oxygen radicals: toxic derivative of oxygen produced in cells as a side product during metabolism.

Pluripotent cell: a cell that is capable of differentiating into (i.e., becoming) many different cell types.

Protein: a polymer made up of typically a few hundred amino acids. Each protein has a unique three-dimensional structure and assumes a function determined by the sequence of the amino acids.

Recombination: process by which two DNA molecules containing regions of related sequence are broken and rejoined in those regions to generate hybrid molecules.

Ribosome: the protein–RNA complexes that make the proteins encoded by genes.

Ribosylation: *see* **ADP-ribosylation**.

Senescence: late stages of an individual's life span. *See also:* **Cellular senescence**.

Silencing: *see* **Gene silencing**.

SIR (Silent information regulator) proteins: proteins that carry out gene silencing.

Single nucleotide polymorphisms (SNPs): small differences in the DNA sequences of individuals within a population.

Stem cells: cells formed early in an organism's development from which many cell types are subsequently generated. *See also:* **Pluripotent cell** and **Differentiation**.

Survival: continuation of life in an individual within a species.

Telomeres: structures at the ends of **chromosomes** that ensure proper maintenance and copying of those **chromosomes**.

Index